CALIFORNIA PLANT FAMILIES

# CALIFORNIA PLANT FAMILIES

West of the Sierran Crest and Deserts

## Glenn Keator

*Illustrations by Margaret J. Steunenberg*

UNIVERSITY OF CALIFORNIA PRESS
Berkeley   Los Angeles   London

The publisher gratefully acknowledges the generous support of
the August and Susan Frugé Endowment Fund in California
Natural History of the University of California Press Foundation.

University of California Press, one of the most distinguished university
presses in the United States, enriches lives around the world by
advancing scholarship in the humanities, social sciences, and natural
sciences. Its activities are supported by the UC Press Foundation and
by philanthropic contributions from individuals and institutions. For
more information, visit www.ucpress.edu.

University of California Press
Berkeley and Los Angeles, California

University of California Press, Ltd.
London, England

© 2009 by The Regents of the University of California

**Library of Congress Cataloging-in-Publication Data**

Keator, Glenn.
  California plant families : west of the Sierran crest and deserts /
Glenn Keator and Margaret J. Steunenberg ; illustrations by Margaret J.
Steunenberg.
      p. cm.
  Includes index.
  ISBN 978-0-520-23709-4 (cloth : alk. paper)—ISBN 978-0-520-25924-9
(pbk. : alk. paper)  1. Plants—California—Identification.  I. Steunenberg,
Margaret J.  II. Title.
  QK149.K34 2009
  581.9794—dc22
                                                            2008055349

Manufactured in the United States of America

18  17  16  15  14  13  12  11  10  09
10  9  8  7  6  5  4  3  2  1

The paper used in this publication meets the minimum requirements
of ANSI/NISO Z39.48-1992 (R 1997)(*Permanence of Paper*).

Cover illustration: *Quercus lobata* (valley oak) and *Phorandendron
villosum* (oak mistletoe). *Atlides halesus* (Great Purple Hairstreak) on
host-plant mistletoe. Transparent watercolor with graphite by
Margaret J. Steunenberg.

# CONTENTS

Preface and Acknowledgments / vii

## Introduction / 1

IMPORTANT PLANT FAMILIES IN CALIFORNIA / 2

HOW PLANT FAMILIES, GENERA, AND SPECIES ARE NAMED / 3

HOW TO USE THIS BOOK / 4

## Key to California Plant Families / 7

## California Plant Family Accounts / 15

ACERACEAE (MAPLE FAMILY) / 16

AGAVACEAE (AGAVE FAMILY) / 18

AIZOACEAE (ICEPLANT FAMILY) / 20

ALLIACEAE (ONION FAMILY) / 21

ANACARDIACEAE (SUMAC FAMILY) / 22

APIACEAE (PARSLEY OR CARROT FAMILY) / 24

APOCYNACEAE (DOGBANE FAMILY) / 28

ARECACEAE (PALM FAMILY) / 32

ASTERACEAE (DAISY, COMPOSITE, OR SUNFLOWER FAMILY) / 34

BERBERIDACEAE (BARBERRY FAMILY) / 42

BETULACEAE (BIRCH FAMILY) / 44

BORAGINACEAE (BORAGE OR FORGET-ME-NOT FAMILY) / 46

BRASSICACEAE (MUSTARD FAMILY) / 48

CACTACEAE (CACTUS FAMILY) / 51

CAMPANULACEAE (BELLFLOWER FAMILY) / 54

CAPPARACEAE (CAPER FAMILY) / 56

CAPRIFOLIACEAE (HONEYSUCKLE FAMILY) / 57

CARYOPHYLLACEAE (PINK FAMILY) / 59

CHENOPODIACEAE (GOOSEFOOT FAMILY) / 61

CONVOLVULACEAE (MORNING GLORY FAMILY) / 63

CORNACEAE (DOGWOOD FAMILY) / 64

CRASSULACEAE (STONECROP OR LIVE-FOREVER FAMILY) / 66

CUCURBITACEAE (GOURD OR CUCUMBER FAMILY) / 68

CUPRESSACEAE (CYPRESS FAMILY) / 70

CYPERACEAE (SEDGE FAMILY) / 72

EPHEDRACEAE (JOINT-FIR FAMILY) / 74

ERICACEAE (HEATHER FAMILY) / 75

EUPHORBIACEAE (SPURGE FAMILY) / 79

FABACEAE (LEGUMINOSAE; PEA OR LEGUME FAMILY) / 81

FAGACEAE (OAK OR BEECH FAMILY) / 84

GENTIANACEAE (GENTIAN FAMILY) / 86

GERANIACEAE (GERANIUM OR CRANESBILL FAMILY) / 88

GROSSULARIACEAE (GOOSEBERRY FAMILY) / 89

HIPPOCASTANACEAE (HORSE-CHESTNUT FAMILY) / 90

HYDROPHYLLACEAE (WATERLEAF FAMILY) / 92

IRIDACEAE (IRIS FAMILY) / 94

JUNCACEAE (RUSH FAMILY) / 96

LAMIACEAE (MINT FAMILY) / 98

LAURACEAE (LAUREL FAMILY) / 101

LILIACEAE (LILY FAMILY) / 103

LOASACEAE (BLAZING STAR FAMILY) / 107

MALVACEAE (MALLOW FAMILY) / 108

MYRTACEAE (MYRTLE FAMILY) / 111

NYCTAGINACEAE (FOUR O'CLOCK FAMILY) / 113

OLEACEAE (OLIVE FAMILY) / 115

ONAGRACEAE (EVENING-PRIMROSE FAMILY) / 116

ORCHIDACEAE (ORCHID FAMILY) / 119

PAPAVERACEAE (POPPY FAMILY) / 122

PINACEAE (PINE FAMILY) / 124

PLATANACEAE (PLANE TREE FAMILY) / 127

POACEAE (GRASS FAMILY) / 128

POLEMONIACEAE (PHLOX FAMILY) / 134

POLYGONACEAE (BUCKWHEAT FAMILY) / 137

PORTULACACEAE (PORTULACA OR PURSLANE FAMILY) / 141

PRIMULACEAE (PRIMROSE FAMILY) / 143

PTERIDACEAE (BRAKE FERN FAMILY) / 145

RANUNCULACEAE (BUTTERCUP OR CROWFOOT FAMILY) / 147

RHAMNACEAE (BUCKTHORN FAMILY) / 150

ROSACEAE (ROSE FAMILY) / 152

RUBIACEAE (MADDER OR COFFEE FAMILY) / 156

RUTACEAE (RUE OR CITRUS FAMILY) / 158

SALICACEAE (WILLOW FAMILY) / 160

SAXIFRAGACEAE (SAXIFRAGE FAMILY) / 162

SCROPHULARIACEAE (FIGWORT OR SNAPDRAGON FAMILY) / 165

SOLANACEAE (POTATO OR NIGHTSHADE FAMILY) / 168

TAXACEAE (YEW FAMILY) / 170

TAXODIACEAE (REDWOOD OR BALD-CYPRESS FAMILY) / 172

THEMIDACEAE (BRODIAEA FAMILY) / 173

VERBENACEAE (VERBENA FAMILY) / 175

VIOLACEAE (VIOLET FAMILY) / 177

Glossary / 179

Common Plant-Names Index / 193

Scientific Plant-Names Index / 205

About the Author and Illustrator / 215

# PREFACE AND ACKNOWLEDGMENTS

When I first undertook this book, it was with the idea of creating a complete guide to California plant families from an earlier version I had worked on many years ago. Because the original material was in bits and pieces, it took some time to assemble them and make them as consistent as possible. That original version contained information on important genera and even species for the common and unusual California families.

But there was controversy over the content of the book and it has now become much more succinct. Reducing the number of families to those most often encountered and the most diverse has resulted in a more cohesive manuscript suitable to the general nature lover and beginning botanist for whom this book was intended. The manuscript has changed vastly and now features a more linear and logical presentation of the material while still retaining the original important genera and species.

Perhaps some readers will be disappointed that I have left out favorite families, but I have carefully assessed the families important throughout the *cismontane* part of this beautiful

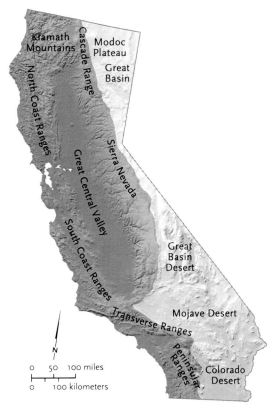

Dark gray indicates cismontane California, the region covered in this book.

state, from the Oregon border to Baja California. Although that term may sound technical, it simply refers to the region that holds the major portion of our great flora, a region that, because of its uniqueness, is called the California Floristic Province.

My thanks go to all the people involved in producing this final product. Among them are editors and assistants at U.C. Press including Doris Kretschmer, Jenny Wapner, and Matthew Winfield, all of whom have helped guide me through the review process.

Peg Steunenberg, my artist, has done a superb job of drawing the many aspects of plants for each family and has done so in a timely fashion despite a difficult schedule. The book would not be the same without her key contributions. Also, I'd like to acknowledge my good friend Susan Bazell, who has allowed me to reuse line drawings she did for a glossary in another of my books: *Plants of the East Bay Parks*.

Finally, I am grateful to the readers who have made valuable suggestions about the content and shape of the book. The result is a much better and more consistent manuscript.

GLENN KEATOR
*Berkeley, California, 2008*

# Introduction

California is a varied state noted for its diverse topography, geology, plant communities, and native plants. Of the 173 or so native and naturalized vascular plant families, there are in excess of 5,800 species, around 1,000 of them introduced from other parts of the world. That is an impressive figure for an area our size.

How can a naturalist make sense of this tremendous diversity? The first step is to look at the big picture by learning to recognize plant families. Despite the intimidating number of those families, roughly 40 to 50 families contain more than 85 percent of our flora. Learning to recognize these keystone families goes a long ways toward the process of identification and allows you to create a framework for most of the genera and species you are likely to encounter. Although there are many versions of keys to these and other families, learning the major field characteristics (often with the aid of a good hand lens) saves a lot of time and makes the process more enjoyable. Close observation of habit, leaves, flowers, and fruits leads to important information that is useful in other ways. For example, it helps to explain pollination, adaptations to habitats, and life cycles.

This book was written with these ideas in mind. The families I have selected are those 50 important ones plus several others that help define our vegetation. For example, maples (family Aceraceae), California buckeye (family Hippocastanaceae), California bay (family Lauraceae), and alders (family Betulaceae) are such key components of woodlands and forests that I felt compelled to include them even though each family has only a few species.

Because there is already an excellent book on major desert families (*California Desert Flowers: An introduction to families, genera, and species* by Sia and Emil Morhardt, published by the University of California Press), I have tried to avoid delving into detail on desert plants, although it has been impossible to exclude them all. The major families selected for this book characterize the area we refer to as the *California Floristic Province*, which also coincides with a geographical realm known as *cismontane California*. Just what is meant by those terms?

Floristic provinces are assemblages of plants that belong to plant communities typical of regions that have a similar overall climate and geographical coherence. The plants in each province form repeatable associations that show adaptations to their home and differ, sometimes dramatically, from plants from other floristic provinces. The California floristic province is one of the most distinctive in the world and represents one of five major areas with a Mediterranean climate—cool to cold, wet winters and warm to hot, usually dry summers. This province extends from the Rogue River in southwestern Oregon south through California into the northwestern fringe of Baja California. Most of it lies to the west of the main mountain crests of the Klamath Mountains, Sierra Nevada, Transverse Ranges, and Peninsular Ranges, a high backbone that delineates much of California. This region is also referred to as cismontane, that is, the region on the ocean side of the mountain crests. The transmontane region to the east is mainly desert and supports other floristic provinces.

As with other aspects of nature, the lines between provinces is often blurred so that various desert elements and families enter the coastal mountains of Southern California. As a consequence, I have included certain families such as the cactus family (Cactaceae) and spurge family (Euphorbiaceae), which are best represented in our deserts. In this endeavor, I have tried my best to be even handed and to balance important plants from all parts of the state.

## IMPORTANT PLANT FAMILIES IN CALIFORNIA

California has around 173 different vascular plant families, some native and others introduced. The introduced ones may appear to grow on their own in natural habitats. Many of these nonnatives are accidental introductions brought in on bricks or ballast, as contaminants in cultivated crops, on domestic livestock, and by humans. Other plants have "escaped" from gardens and cultivated fields to grow on their own. Many of these are invasive and seriously threaten the diversity of our native flora.

We live in a time of rapid change in the world of biology. It is important to know that the state of defining plant families is changing more dramatically than at any previous time in history. Despite the inconvenience of a classification in a state of flux, think of these changes as an exciting challenge to learn more about the evolutionary relationships of all organisms. Certainly these changes are frustrating to those wishing to learn an unchanging system, but that is not the reality of what is going on.

The reason for this state of affairs is that many important lines of research are creating a more complete picture of how species and genera are related, and which families they belong to. In addition to the more classical lines of inquiry such as external form (morphology), anatomy, chromosome studies, and details of pollen, the growing arsenal of information includes studies of biochemistry involving pigments, poisons, perfumes, proteins, and many other compounds. Add to this the rapidly expanding field of DNA studies that plot the rate of changes in selected genes, and you have a far richer and more detailed story of true relationships. Computer-generated cladograms that display the degree of relatedness of plant groups present an evolutionary scheme displayed as a branched system. These cladograms are used to determine the limits of families and genera.

Consequently, several well-known and widely studied families are now in a state of change. I wish I could say that these changes will be permanent and all you need do is learn the new classifications, but as we continue to learn more and examine a larger array of genes, we will not only refine the now-current system but also make additional changes.

Some long-recognized families have been split into two or more separate families while others have been "lumped" together. Still others are so complex that totally new alignments are being made.

My approach in this book is somewhat conservative; because the current *Jepson Manual* is still the standard reference for workers and students and because so many other local floras follow a similar system for the families, I have continued to use many of the familiar family definitions. But I also alert you, the reader, to changes that are proposed or have been accepted; future field books will eventually reflect these changes.

I have deviated from the older system(s) with the large, important lily family (Liliaceae), which is now considered to consist of many separate, sometimes unrelated families. Although I have not treated all of these splinter families here, I am describing some of the major ones that are easily recognized: the agave family (Agavaceae), onion family (Alliaceae), and brodiaea family (Themidaceae).

## HOW PLANT FAMILIES, GENERA, AND SPECIES ARE NAMED

All plants have names, and it is the name we turn to first when we want to learn more about a particular group. You will find that plants bear two kinds of names: trivial or common names that are in everyday use by the average person, and scientific or Latin names, that are used throughout the horticultural and botanical world. There are advantages and disadvantages to both kinds of names, but for greater precision, scientific names are preferred. Common names are not always standardized, and many plants and plant families—for example, Fagaceae, aka the oak or beech family—have more than one common name. Common names may also allude to relationships that do not exist. For example, the evening-primrose family (Onagraceae) and primrose family (Primulaceae) are not at all closely related; evening-primroses and their relatives belong to a different evolutionary line. Perhaps the first people to notice this beautiful family were struck by the showy flowers that they imagined looked like oversized primroses, but botanically, the two families differ by many traits (turn to pp. ooo for a description). When these sorts of common names have become so embedded in the language that they are permanent, the name is hyphenated to indicate that it represents a special combination. So we have corn-lily (*Veratrum* spp.) for a perennial wildflower that belongs to the lily family but is not a true lily (*Lilium* spp.), and Douglas-fir (*Pseudotsuga menziesii*) for a tree that despite its very different cones and bark reminded someone of a fir (*Abies* spp.).

Not every plant or plant family has a well-established common name, but since people want a common name, I have tried to provide them when possible. Not everyone will agree with my choice of common names, and some will disagree with my use of one common name over another. Where there is more than one well-known common name, I have included it.

Scientific names are based on carefully crafted rules for naming, and are in a latinized form. Because scientific names can be recognized by scientists throughout the world regardless of the language they speak, the names give a real sense of permanence and are immediately recognizable anywhere.

## PLANT FAMILIES

Because this book focuses first and foremost on families, I will talk first about the rules for scientific family names. (Families are larger than genus and species, often embracing several different genera and many species, although some families are very small.) All family names end in -aceae and are based on a *type genus:* for example, we have Rosaceae (rose family) based on the type genus *Rosa;* Liliaceae (lily family) named for the type genus *Lilium;* and Orchidaceae (orchid family) from the type genus *Orchis.* Those examples are based on names that are cognates of common English names, but many family names are not recognizable, such as Ranunculaceae (buttercup family), Rhamnaceae (buckthorn family), and Scrophulariaceae (figwort family). Understanding the derivation of these names makes learning them much more enjoyable. Ranunculaceae refers to *little frogs*, because many buttercups live in wet areas, the habitats for frogs. Scrophulariaceae is named for the type genus *Scrophularia*, based on the belief that it cured the skin disease known as scrofula.

It is also important to understand that scientific names may change as more is learned about a particular group. For example, there is considerable controversy over what belongs to the lily family (Liliaceae). Research indicates that there are several separate evolutionary lines in this family if you define it to include the broadest possible concept. Each line is often given a separate family name. But if you look at the bigger picture and choose to focus on the common ancestry to these evolutionary lines, you may prefer to lump all the species together into one very large, inclusive family.

I have usually followed the family concepts given in the *Jepson Manual* but as commented on above, new DNA research is turning many families topsy-turvy. In short, evidence from this research is causing an almost unprecedented reorganization of families. The Liliaceae mentioned above is now sliced into more than 16 separate families (not all of these are Californian). This reorganization is highly relevant to those studying evolutionary relationships but it raises havoc with conventional books for the amateur botanist and nature lover.

PLANT GENERA

Besides families, this book covers many important genera and even some species. After family comes genus (plural genera). A genus consists of closely related kinds of plants. For example, among wildflowers we have violets (genus *Viola*), daisies (genus *Erigeron*), Indian paintbrushes (genus *Castilleja*), buttercups (genus *Ranunculus*). Some genera have only one species; others have dozens or even hundreds. The kinds of violets or daisies are what we call species (singular and plural are the same). California is blessed with many species of violets, including *V. douglasii* (Douglas's violet), *V. pedunculata* (wild pansy), *V. adunca* (dog violet), *V. macloskeyi* (white meadow violet), *V. ocellata* (western heartsease), and many more.

PLANT SPECIES

Each scientific species name consists of two parts: a genus name, given first and capitalized; and a specific epithet, given second and starting with a lowercase letter. (Think of how most people use two names to identify themselves.)

Both genus name and specific epithet are underlined or italicized. Although the initial reaction of the novice may be that scientific names are impossibly difficult, in fact English speakers have a decided advantage since many of the latinized names have cognates in English. A few examples of recognizable scientific names follow: *Lilium maritimum* (coast lily)—the genus name is the Latin version of lily, followed by a word that means coastal (maritime). *Viola purpurea* (pine or oak violet). The name *Viola* alludes to the violet color of many flowers in the genus, although this particular species happens to be yellow; *purpurea* means *purple*, perhaps because the underside of the mature leaves is purple. *Delphinium nudicaule* (scarlet larkspur): *Delphinium* comes from the Latin word for dolphin and is a cognate of our English word—larkspurs have a sleek, streamlined outline that is reminiscent of the body shape of dolphins; *nudi* means *naked* (nude), *caule* stem (think of the word *cauliflower*, meaning *stem flower*).

HOW TO USE THIS BOOK

If you already know the name of the plant but are curious about the family it belongs to or are simply trying to organize your knowledge better, turn directly to the main section of the book, where each family is listed alphabetically by its scientific name. If you know only the common name of the family, consult the index. If you have a plant to identify and have no idea what it is, refer to the family key (below).

CALIFORNIA PLANT FAMILY ACCOUNTS

Each family entry starts with a short statement about recognition at a glance—what to look for as a first step to identifying the family. When this does not suffice—or if you are anxious to learn more—a series of descriptions expands on other family traits including the habit, leaves, branches, flower shape and arrangement, flower parts—numbers of sepals, petals, stamens, and pistils—and fruits. The term *habit* describes the form of a plant, such as shrub, tree, herbaceous perennial, annual, or bulb.

Because many families look alike or are closely related, there is a section dealing with similar-looking families (which are not always closely related despite a superficial resemblance), with quick suggestions to distinguish them from the family you are learning.

The *statistics* entry alerts you to the distribution, habitat preferences, size, and economic uses of the family.

Finally, there is a section on California genera and species. For small families, I have tried to be as inclusive as possible so that, for example, you will learn about all four tree species that belong to the maple family (Aceraceae). For large families with dozens to hundreds of species, a comprehensive breakdown of the family is beyond the scope of the book, but I have included a fair sampling of genera and species that I consider typical, common, or interesting. These examples are presented under logical categories that act as a guide or simplified key to separate the given genera.

## CALIFORNIA PLANT FAMILY KEY

The key is designed to minimize technical features, although many details such as types of fruits and the ways anthers shed their pollen are necessarily mentioned. A good 10x hand lens is a valuable and necessary adjunct to see many of these traits. I often include more than one trait for each step in the key but sometimes the second trait given is not necessary to successfully use the key. (Second or third traits at each step are there to help confirm your decision.)

Although I have attempted to take into consideration exceptions to the general family characteristics, there will doubtless be some that I have missed; the key should work to identify a plant to family most of the time.

If you have a plant to identify, you will want to use the simplified key and carefully follow the choices at each step. To start keying, choose one of the three groups—conifers, monocots and monocotlike plants, or dicots—then turn to that group and proceed from there. Most steps in the key have two choices but a few have three or rarely more.

If you are new to keying, here are some pointers to bear in mind:

- Keys are imperfect because nature is not always consistent.
- Be sure to note as many features about the plant as possible if you do not have a fresh specimen in hand. You will want to note the leaf shape, arrangement, and form (simple or compound), the habit of the plant (herb, shrub, or tree), the arrangement of the flowers, and the details of the flowers and their parts, including shape and any special features. If your specimen has fruits or seeds, those should also be noted.
- Write down the choices you have made so that when a particular choice is not clear you can backtrack if the family you choose does not fit the description of your plant.
- Try working the key backwards, starting with a family whose name you know. By following the steps that lead to the family name, you will learn more about how the key works.
- Practice makes perfect, and the more repetitions you do, the easier the process becomes.
- Consult the glossary at the back of the book for terms that are not familiar. Learning these terms is like learning a new language.
- Consult the illustrations of flower parts located in the glossary.

As you peruse family traits you may find certain unfamiliar terms. Technical terms are useful when they replace an otherwise sentence-long description. The glossary will help make sense of these terms. Be sure to study the line drawings to see how the language translates visually.

Similar-looking families may seem confusing to separate. (Examples include the borage and waterleaf families, the morning glory and nightshade families, and the onion and brodiaea families.) In such cases, check the description of the look-alike family. I comment on families that might be confused with one another and compare them.

Finally, note that no single trait ever serves as a sure means of recognizing a family; rather, you need to look at a combination of traits. For example, knowing that the Onagraceae (evening-primrose family) is characterized by an inferior ovary does not allow you to separate it from other families with inferior ovaries. But when you note that your plant also has four petals, four sepals, four or eight stamens, and a capsule-type fruit, you can be confident that the plant belongs to this family and no other.

Families sometimes differ in their variability. For example, the Asteraceae (daisy family) and Apiaceae (parsley family) are intuitively easy to recognize because of the appearance of their flowers: Daisies have many flowers packed together in a head that resembles a single flower; parsleys bear many tiny flowers in compound umbels. By contrast, the Rosaceae (rose family) varies in just about every feature from leaf design to flower arrangement, from plant habit to stamen number, and from habitat to ovary position. Yet, even this large family offers important clues to its identity such as the presence of a hypanthium, a single-rose-like flower design, specific types of fruits, the frequent presence of stipules on the leaves, and sepal-like bracts that alternate with the true sepals on flowers of the nonwoody species.

Remember: Practice makes keying go faster and more smoothly. Use this book often and you will grow more proficient at keying and find satisfaction in your new skills.

# Key to California Plant Families

1 Seed-bearing plants with seed cones or flowers, go to 2
1' Spore-bearing plants without seeds, go to Pteridaceae (brake fern family)

2 Cone-bearing seed plants, go to Group 1
2' Flowering plants with sepals or petals in threes, go to Group 2
2" Flowering plants with sepals or petals in fours, fives, or multiples, go to Group 3

Group 1. CONIFERS (PINE, CYPRESS, YEW, AND REDWOOD FAMILIES)
1 Shrubs with jointed green twigs and leaves reduced to scales....**Ephedraceae (joint-fir family)**
1' Trees (one species a needle-bearing shrub) without these features, go to 2

2 Mature leaves needlelike, go to 3
2' Mature leaves scalelike (if needlelike, borne in pairs or whorls)....**Cupressaceae (cypress family)**

3 Needles shed with twigs....**Taxodiaceae (redwood or bald cypress family)**
3' Needles shed separately or on tiny spur shoots, go to 4

4 Trees monoecious; seed cones papery or woody....**Pinaceae (pine family)**
4' Trees dioecious; seed cones fleshy....**Taxaceae (yew family)**

Group 2. MONOCOTS AND PRIMITIVE DICOTS (SEVERAL DIFFERENT FAMILIES)
1 Leaf-vein pattern branched or netlike, go to 2
1' Leaf-vein pattern parallel, go to 7

2 Woody shrubs and trees, go to 3
2' Herbaceous plants, go to 5

3   Leaves simple, strongly aromatic. . . .**Lauraceae (laurel family)**
3'   Leaves compound or deeply divided, not aromatic, go to 4

4   Shrubs or ground covers; leaves compound. . . .**Berberidaceae (barberry family)**
4'   Single-trunked trees; leaves deeply palmately lobed. . . .**Arecaceae (palm family)**

5   Single flower borne just above the leaf whorl. . . .**Liliaceae, genus *Trillium***
5'   Flower(s) not arranged like this; leaves are not whorled, go to 6

6   Leaves simple. . . .**Polygonaceae (buckwheat family)**
6'   Leaves compound. . . .**Berberidaceae (barberry family)**

7   Grasslike plants with small, mostly greenish or brownish flowers, go to 8
7'   Plants *not* both grasslike *and* with small, greenish flowers, go to 9

8   Perianth obvious (use hand lens). . . .**Juncaceae (rush family)**
8'   Perianth clearly missing or highly modified and thus not obvious, go to 9

9   Stems solid, often three-sided; leaves channeled. . . .**Cyperaceae (sedge family)**
9'   Stems usually hollow, usually round; leaves seldom channeled. . . .**Poaceae (grass family)**

10   Flowers irregular; lower petal enlarged into a lip. . . .**Orchidaceae (orchid family)**
10'   Flowers regular or nearly so; no obvious lip, go to 11

11   Leaves sword-shaped and equitant; flower buds protected inside pairs of bracts. . . .**Iridaceae (iris family)**
11'   Leaves seldom sword-shaped and equitant; flower buds not enclosed inside pairs of bracts, go to 12

12   Leaves with strong, tough fibers; flowers usually in very large inflorescences several feet long. . . .**Agavaceae (agave family)**
12'   Leaves not strongly fibrous; flowers not in uncommonly large inflorescences, go to 13

13   Flowers arranged in bracted umbels, go to 14
13'   Flowers arranged in some other way. . . .**Liliaceae (lily family)**

14   Leaves strongly onion-scented; tepals seldom joined to form a tube. . . .**Alliaceae (onion family)**
14'   Leaves not onion-scented; tepals usually joined to form a tube. . . .**Themidaceae (brodiaea family)**

## Group 3. MOST DICOTS (THE BULK OF THE FLOWERING FAMILIES)

1   Flowers both tiny and petalless, go to 2
    Flowers (apparently) have petals, go to 14

2   Flowers in dense heads surrounded by sepal-like bracts. . . .**Asteraceae (ragweed subtribe of composite or daisy family)**
2'   Flowers not in dense heads as above, go to 3

3   Leaves usually covered with dense hairs or mealy scales, go to 4
3'   Leaves not like this, go to 5

4   Ovary not strongly lobed; style one. . . .**Chenopodiaceae (goosefoot family)**
4'   Ovary strongly three-lobed; styles three. . . .**Euphorbiaceae (spurge family)**

5   Shrubs or trees, go to 6
5'  Herbaceous plants, go to 13

6   Leaves compound, go to 7
6'  Leaves simple (may be lobed), go to 8

7   Fruits a double-winged samara. . . .**Aceraceae (*Acer negundo*, maple family)**
7'  Fruits a single-winged samara. . . .**Oleaceae (olive family; *Fraxinus* spp.)**

8   Leaves deeply palmately lobed. . . .**Platanaceae (plane tree family)**
8'  Leaves not palmately lobed, go to 9

9   Leaves thin and deciduous, go to 10
9'  Leaves tough and evergreen. . . .**Fagaceae (oak family)**

10  Male flowers in catkins, go to 11
10' Male flowers not in catkins, *or* flowers bisexual. . . .**Rhamnaceae (buckthorn family)**

11  Leaves lobed; fruit an acorn in a scaly cup. . . .**Fagaceae (oak family)**
11' Leaves not lobed; fruits not acorns, go to 12

12  Seeds hairless *and* fruits are tiny winged achenes or nuts; plants monoecious. . . .**Betulaceae (birch family)**
12' Seeds with dense hairs and borne in capsules; plants dioecious. . . .**Salicaceae (willow family)**

13  Leaves have stipules; flowers have a hypanthium. . . .**Rosaceae (rose family)**
13' Leaves lack stipules; flowers lack a hypanthium. . . .**Ranunculaceae (buttercup family)**

14  Flowers neither sweetpealike nor daisylike, go to 15
14' Flowers in daisylike heads resembling single flowers. . . .**Asteraceae (daisy or sunflower family)**

14" Flowers sweetpealike. . . .**Fabaceae (pea family)**

15  "Petals" actually colorful nectar-secreting glands surrounding a cup of minute flowers. . . .**Euphorbiaceae (spurge family)**
15' True petals present, go to 16

16  Stamens numerous, go to 17
16' Stamens 10 or fewer, go to 25

17  Ovary inferior, go to 18
17' Ovary superior, go to 22

18  Stems fleshy and spiny; flowers with multiple petals. . . .**Cactaceae (cactus family)**
18' Stems not fleshy; flowers with a single row of petals, go to 19

19  Flowers with a hypanthium; leaves seldom fragrant. . . .**Rosaceae (rose family)**
19' Flowers lack a hypanthium; leaves sometimes fragrant, go to 20

20  Shrubs or trees with highly fragrant leaves. . . .**Myrtaceae (myrtle family)**
20' Herbaceous or subwoody plants; leaves not fragrant, go to 21

21  Leaves fleshy; petals often numerous. . . .**Aizoaceae (iceplant family)**
21' Leaves not fleshy; petals 10 or fewer. . . .**Loasaceae (blazing star family)**

22  Stamens separate, go to 23
22' Stamens fused to form a hollow tube....**Malvaceae (mallow family)**

23  Flowers have a hypanthium (look carefully; hypanthium may be small)....**Rosaceae (rose family)**
23' Flowers lack a hypanthium, go to 24

24  Several to many separate pistils....**Ranunculaceae (buttercup family)**
24' A single compound pistil....**Papaveraceae (poppy family)**

25  Ovary (apparently) inferior, go to 26
25' Ovary superior, go to 36

26  Flowers in umbels of umbels *or* umbels of heads....**Apiaceae (parsley family)**
26' Flowers in other arrangements, go to 27

27  Vines *or* plants with milky sap, go to 28
27' Not vines; milky sap absent, go to 29

28  Vines with tendrils; flowers unisexual....**Cucurbitaceae (cucumber or squash family)**
28' Self-supporting plants with milky sap; flowers mostly bisexual....**Campanulaceae (bellflower family)**

29  Herbaceous plants, go to 30
29' Woody plants, go to 32

30  Ovary not truly inferior; enclosed in a saclike hypanthium....**Nyctaginaceae (four o'clock family)**
30' Ovary truly inferior, go to 31

31  Flowers usually have a hypanthium developed above the ovary....**Onagraceae (evening-primrose family)**
31' Flowers lack a hypanthium....**Rubiaceae (madder family)**

32  Sepals and petals both colored; hypanthium present....**Grossulariaceae (gooseberry family)**
32' Sepals greenish, often inconspicuous; hypanthium missing, go to 33

33  Leaves have stipules (look closely)....**Rubiaceae (madder family;** *Cephalanthus occidentalis*)
33' Leaves lack stipules, go to 34

34  Petals four; flowers in cymes or heads....**Cornaceae (dogwood family)**
34' Petals five; flowers in varied arrangements, go to 35

35  Leaves opposite....**Caprifoliaceae (honeysuckle family)**
35' Leaves alternate....**Ericaceae (heather family)**

36  Petals separate, not joined, go to 37
36' Petals partly joined, falling together as a unit, go to 66

37  Flowers irregular, go to 38
37' Flowers regular, go to 44

38  Trees or shrubs, go to 39
38' Herbaceous plants, go to 39

39  Leaves often ill-smelling; flowers with six stamens....**Capparaceae (caper family)**
39' Leaves not ill-smelling; flowers with five stamens....**Violaceae (violet family)**

40  Trees with palmately compound leaves. . . .**Hippocastanaceae (horse-chestnut family)**
40′  Shrubs with simple or compound leaves, go to 41

41  Ovary with beaklike styles, five-chambered. . . .**Geraniaceae (geranium family, genus *Pelargonium*)**
41′  Ovary lacks beaklike styles and is one- or two-chambered, go to 42

42  Stamens five or 10; fruit a one-chambered legume. . . .**Fabaceae (pea family)**
42′  Stamens six; fruit not a legume, go to 43

43  Leaves palmately divided or compound. . . .**Capparaceae (caper family)**
43′  Leaves not palmate. . . .**Brassicaceae (mustard family)**

44  Herbaceous plants, go to 45
44′  Woody, spineless plants, go to 61
44″  Woody, spiny plants. . . .**Rhamnaceae (buckthorn family)**

45  Leaves fleshy, go to 46
45′  Leaves not especially fleshy, go to 48

46  One pistil per flower; two (usually) or four sepals, go to 47
46′  Five separate or only partly joined pistils; five sepals. . . .**Crassulaceae (stonecrop family)**

47  Usually two sepals (sometimes an indefinite number) and a single compound pistil. . . .**Portulacaceae (portulaca family)**
47′  Four sepals and four petals. . . .**Brassicaceae (mustard family)**

48  Leaves dissected or compound, go to 48
48′  Leaves simple (may be lobed), go to 53

49  Flowers have a hypanthium; sepal-like bracts alternate with sepals. . . .**Rosaceae (rose family)**
49′  Flowers lack a hypanthium; sepal-like bracts missing, go to 50

50  Woody plants with strongly scented leaves. . . .**Rutaceae (rue family; *Ptelea crenulata*)**
50′  Herbaceous plants with unscented to lightly scented leaves, go to 51

51  Petals and sepals four; styles not forming a beak, go to 52
51′  Petals and sepals five; styles forming a beak. . . .**Geraniaceae (geranium family)**

52  Leaves palmately divided or compound. . . .**Capparaceae (caper family)**
52′  Leaves not palmate. . . .**Brassicaceae (mustard family)**

53  Ovaries look inferior because they form a bulge inside the hypanthium. . . .**Nyctaginaceae (four o'clock family)**
53′  Ovaries look superior, go to 54

54  Leaves opposite. . . .**Caryophyllaceae (pink family)**
54′  Leaves generally alternate or basal, go to 55

55  Leaves have stipules, go to 56
55′  Leaves usually lack stipules, go to 57

56  Stipules form papery sheaths; styles are not beaked. . . .**Polygonaceae (buckwheat family)**
56′  Stipules are green and not sheathing; styles form a beak. . . .**Geraniaceae (geranium family)**

57  No distinction between sepals and petals. . . .**Polygonaceae (buckwheat family)**
57' Sepals and petals are obviously different, go to 58

58  Hypanthium present; often two separate or partly joined pistils. . . .**Saxifragaceae (saxifrage family)**
58' Hypanthium is missing; one single pistil, go to 59

59  Woody plants with highly aromatic leaves. . . .**Rutaceae (rue family)**
59' Herbaceous plants with mostly unscented leaves, go to 60

60  Four petals; four sepals. . . .**Brassicaceae (mustard family)**
60' Six petals and three sepals. . . .**Papaveraceae (poppy family)**

61  Leaves compound, go to 62
61' Leaves simple, go to 63

62  Leaves pinnately compound; fruit a legume. . . .**Fabaceae (pea family)**
62' Leaves trifoliate; fruit a berrylike drupe. . . .**Anacardiaceae (sumac family)**

63  Leaves opposite, go to 64
63' Leaves alternate, go to 65

64  Leaves palmately lobed; fruits are double-winged samaras. . . .**Aceraceae (maple family)**
64' Leaves not lobed; fruit is a capsule. . . .**Rhamnaceae (buckthorn family)**

65  Sepals colored; ovary three-chambered. . . .**Rhamnaceae (buckthorn family)**
65' Sepals green; ovary not three-chambered. . . .**Anacardiaceae (sumac family)**

66  Flowers regular, symmetrical, go to 67
66' Flowers clearly irregular, go to 79

67  Leaves and stems have milky sap, go to 68
67' Leaves and stems lack milky sap, go to 69

68  Two ovaries per flower; many seeds per ovary. . . .**Apocynaceae (dogbane family)**
68' One ovary per flower; few seeds per ovary. . . .**Convolvulaceae (morning glory family)**

69  Flowers arranged mostly in coiled clusters that unfurl as flowers open, go to 70
69' Flowers not arranged this way, go to 71

70  Style single, not divided; ovary four-lobed. . . .**Boraginaceae (borage or forget-me-not family)**
70' Style usually partway split in two; ovary two-chambered. . . .**Hydrophyllaceae (waterleaf family)**

71  Woody plants, go to 72
71' Herbaceous plants, go to 74

72  Stamens two per flower. . . .**Oleaceae (olive family)**
72' Stamens four to 10 per flower, go to 73

73  Petals not pleated in bud; stamens attached to a disc. . . .**Ericaceae (heather family)**
73' Petals pleated in bud; stamens attached to petals. . . .**Solanaceae (nightshade family)**

74  Three stigma lobes and a three-chambered ovary. . . .**Polemoniaceae (phlox family)**
74' One or two stigma lobes and a one-, two-, or five-chambered ovary, go to 75

75 Petals pleated or folded in bud, go to 76
75' Petals not pleated or folded in bud, go to 77

76 Petals often bear appendages; ovary is incompletely two-chambered....**Gentianaceae (gentian family)**
76' Petals lack appendages; ovary is two- or five-chambered....**Solanaceae (nightshade family)**

77 Stamens line up opposite petals; seeds borne on a central stalk inside the ovary....**Primulaceae (primrose family)**
77' Stamens line up alternating with petals; seeds are not on a central stalk, go to 78

78 Styles usually two-forked....**Hydrophyllaceae (waterleaf family)**
78' Styles single, not forked....**Scrophulariaceae (figwort family, genus *Verbascum*)**

79 Leaves often scented; ovary divided into four lobes, go to 80
79' Leaves seldom scented; ovary partly or fully two-chambered....**Scrophulariaceae (figwort family)**

80 Leaves not mint- or sage-scented; flowers slightly irregular....**Verbenaceae (verbena family)**
80' Leaves (usually) strongly mint- or sage-scented; flowers (mostly) two-lipped....**Lamiaceae (mint family)**

# California Plant Family Accounts

# ACERACEAE (Maple Family)

**RECOGNITION AT A GLANCE** In California, trees with opposite, palmately lobed or pinnately compound leaves and double-winged samaras in fruit.

**VEGETATIVE FEATURES** Deciduous shrubs or trees with opposite twigs and branches; leaves are often palmately lobed, compound, or veined.

**FLOWERS** Small, yellow, greenish, or reddish, and often clustered in chainlike racemes or catkins.

**FLOWER PARTS** 5-lobed calyx, (sometimes missing) separate petals, 8 stamens, and a single pistil with a superior, 2-chambered ovary with 2 styles and 2 wings.

**FRUITS** Winged achenes (samaras).

**RELATED OR SIMILAR-LOOKING FAMILIES** Current research places the maple family in the mainly tropical Sapindaceae (soapberry family). By the old definition of that family, there were no California natives.

Hippocastanaceae (horse-chestnut family) has palmately compound leaves, spikelike clusters of showy flowers, and leathery capsules containing a single seed. It too is currently considered to belong to the Sapindaceae.

**STATISTICS** Around 120 species across the cooler parts of the northern hemisphere and especially diverse in eastern Asia. Many species live in mountains or near watercourses. The maples (*Acer* spp.) give us many garden ornamentals. Maple sugar and syrup come from the eastern North American sugar maple (*A. saccharum*).

## CALIFORNIA GENERA AND SPECIES

The region has a single genus, *Acer* (maple), with four species.

*Acer macrophyllum* (bigleaf maple; see fig. a) is a large tree with large, 3- to 5-lobed leaves and pale yellow flowers. Its leaves often turn yellow in fall. It lives in riparian woodlands up to 5,000 feet.

*A. negundo* var. *californicum* (box elder; see fig. b) is a large, dioecious tree with pinnately compound leaves of 3 to 5 leaflets. The petalless flowers are in hanging catkins, the males with a pinkish tint. It is widely scattered in riparian woodlands.

*A. circinatum* (vine maple; see fig. c) is a small tree or large, multi-trunked shrub with small, multilobed leaves and tiny dull red and white flowers. Its leaves turn scarlet in fall. It occurs in the forests of northwestern California.

*A. glabrum* and varieties (Sierra or mountain maple) is a large, multi-trunked shrub with small, glossy, few-lobed leaves and hanging trusses of pale yellow-green flowers. It lives in the upper montane and subalpine zones.

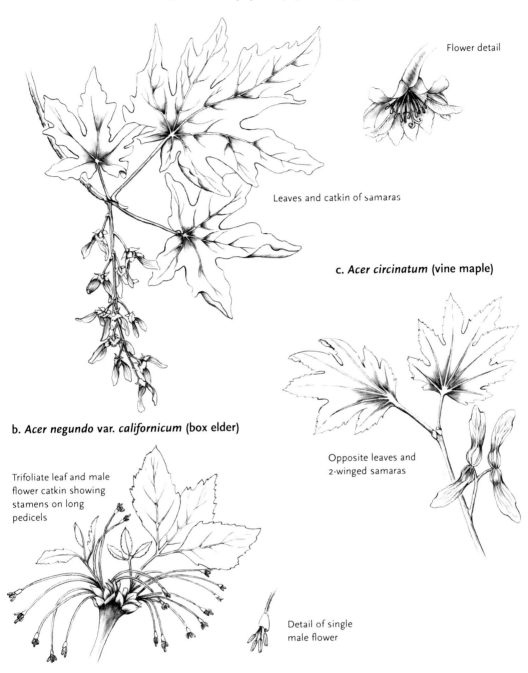

# AGAVACEAE (Agave Family)

**RECOGNITION AT A GLANCE** Bold, rosetted plants with enormous, tough, fibrous leaves and huge panicles of bell-shaped or tubular flowers.

**VEGETATIVE FEATURES** Woody monocots (underground rhizomes to aboveground trunks) with large rosettes of simple, linear to ovate, toughly fibrous, fleshy leaves. Leaves often have spiny teeth.

**FLOWERS** Large, showy, cream-colored, reddish, or yellow and either bell-shaped or tubular. The flowers are arranged in panicles several feet long.

**FLOWER PARTS** 2 rows of 6 fleshy tepals, 6 stamens, and a single pistil with an inferior or superior, 3-lobed ovary.

**FRUITS** Fleshy or woody, 3-lobed capsules containing 6 rows of seeds.

**RELATED OR SIMILAR-LOOKING FAMILIES** The agave family has been variously formulated by different authors at different times. It has sometimes been included in the lily family (Liliaceae). Its broad defintion includes most large monocots with fibrous leaves and large flower clusters but currently the family has been subdivided. Note that the genus *Nolina* (bear-grass), once considered to belong to this family, is now placed in the lily-of-the-valley family (Convallariaceae).

**STATISTICS** Around 400 species concentrated in the dry and desert areas of North and Central America. Besides providing several garden ornamentals for xeriscapes, fibers are often extracted from yucca rhizomes and agave leaves; the best known is sisal from *Agave sisalana*. Some Mexican agaves also provide a sugary sap that is fermented to produce pulque and distilled to make tequila.

## CALIFORNIA GENERA AND SPECIES

The region has three native genera and seven species.

*Agave* (maguey, century plant) has leaves usually lined with vicious, recurved spines, and massive inflorescences of tubular yellow to cream-colored flowers.

*A. shawii* is restricted to the southernmost coastal strip in San Diego County and extends south well into Baja California.

*A. deserti* is common in rocky slopes of the Sonoran Desert.

*A. utahensis* is rare in the mountains of the eastern Mojave Desert and generally lives on limestone outcrops. It is much smaller than the other two species with inflorescences reaching only a foot or two high.

*Hesperoyucca whipplei* (our lord's candle, chaparral yucca; see fig.) has narrow, rigid leaves; large panicles of creamy, bell-shaped flowers (sometimes tinted purple); and blooms once then dies. This monocarpic way of life distinguishes it from the true yuccas.

*Yucca* (yucca, Joshua tree, and other names) The true yuccas bloom many times during their lives.

*Y. baccata* (banana yucca) makes colonizing clumps of basal leaves and has narrow creamy bells flushed red.

*Y. schidigera* (Spanish dagger) produces a sometimes branched trunk and massive panicles of creamy bells.

*Y. brevifolia* (Joshua tree) is a multibranched tree with relatively short leaves and tight racemes of creamy flowers with a greenish tint. It is restricted to the Mojave Desert.

### *Hesperoyucca whipplei* (chaparral yucca or our lord's candle)

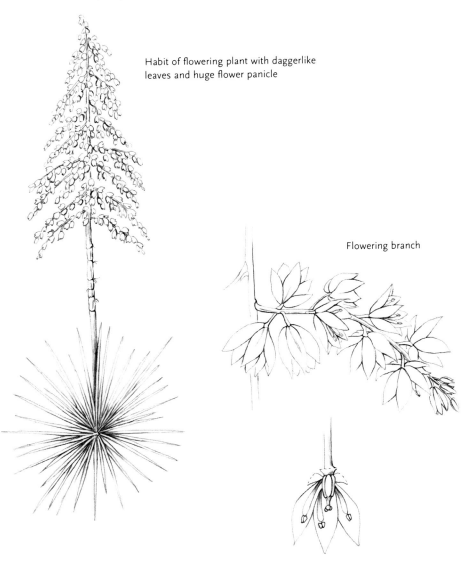

Habit of flowering plant with daggerlike leaves and huge flower panicle

Flowering branch

Flower detail with 2 tepals and stamens removed for clarity

# AIZOACEAE (Iceplant Family)

**RECOGNITION AT A GLANCE** Ground covers and bushy plants with succulent leaves and showy flowers featuring (usually) numerous petals and stamens.

**VEGETATIVE FEATURES** Herbaceous to woody ground covers and perennials with simple, very fleshy, alternate to opposite leaves (sometimes covered with crystal-like glands).

**FLOWERS** Usually large, showy, and white or various bright colors with a starburst pattern.

**FLOWER PARTS** Usually 5 fleshy sepals, (usually) numerous narrow petals, numerous stamens, and a single pistil with an inferior to occasionally superior, multi-chambered ovary.

**FRUITS** Fleshy, berrylike fruits with numerous seeds.

**RELATED OR SIMILAR-LOOKING FAMILIES** This succulent family stands apart from others such as the Crassulaceae (stonecrop family) and Portulacaceae (portulaca family) by its numerous narrow petals and multiple stamens. In addition, the flowers of Crassulaceae usually have 5 separate to partly joined pistils while most flowers in the Portulacaceae feature 2 sepals and 5 petals. There is also a strong similarity in flower design between the Aizoaceae and Cactaceae (cactus family). Cacti differ by having highly modified fleshy stems and clusters of leaves modified into spines whereas Aizoaceae has fleshy leaves and no spines.

**STATISTICS** 2,500 species mostly native to the southern hemisphere, with the greatest diversity in South Africa. The family is noted for several invasive species as well as a plethora of ground covers with brilliantly colored flowers that are especially popular in coastal gardens of central and southern California.

## CALIFORNIA GENERA AND SPECIES

The region has two native genera and species, and nine introduced genera with 11 species.

### NATIVE GENERA

*Sesuvium verrucosum* (western sea-purslane) has a half-inferior ovary and pairs of leaves of equal length. It lives on the edge of salt flats.

*Trianthema portulacastrum* (horse-purslane) has a superior ovary and pairs of leaves of unequal length. It occurs in dry flats in the San Joaquin Valley and deserts.

### NONNATIVE GENERA

*Carpobrotus chilensis* (iceplant) was once believed to be native to our coastal bluffs and dunes but research indicates that it came from South Africa. In addition to the magenta-flowered iceplant noted above, *C. edulis* (hottentot-fig; see fig.) is an invasive coastal ground cover with large cream-colored or magenta blossoms. It crowds out most native plants and is a common ground cover along southern California freeways.

*Mesembryanthemum* (true iceplants) are common along the south coast, especially on salty soils. *M. crystallinum* dramatically illustrates the common name: Its stems, leaves, and sepals are covered with excrescences that resemble ice crystals.

*Conocosia pugioniformis* features narrow leaves and bright yellow flowers. It is found along the south-central and south coasts.

**Carpobrotus edulis (hottentot-fig or iceplant)**

Fleshy leaves and flower with 5 sepals, numerous petals, and numerous stamens

# ALLIACEAE (Onion Family)

**RECOGNITION AT A GLANCE** Onion-scented bulbs with flowers arranged in bracted umbels.

**VEGETATIVE FEATURES** Underground, scaly bulbs send up 1 to 4, long, strap-shaped, onion-scented leaves that often wither by flowering time. Leaves may be flat and linear, sickle-shaped, or hollow and tubular.

**FLOWERS** Small, often star-shaped, and are borne in open, bracted umbels.

**FLOWER PARTS** 6 similar, usually separate tepals in 2 rows; 6 stamens; and a single pistil with a superior, 3-lobed ovary.

**FRUITS** 3-chambered capsules.

**RELATED OR SIMILAR-LOOKING FAMILIES** Over the past centuries, this group has been placed in many other families including Amaryllidaceae (the amaryllis family) and Liliaceae (lily family). (For a more complete discussion of the recent fate of the lilies, see the Liliaceae account.) The combination of flowers with 6 tepals, flowers arranged in bracted umbels, and a strong onion odor separate this family readily from others. The only other California family likely to be confused is the Themidaceae (brodiaea family), whose flowers often have a petal tube and whose leaves lack an onion odor. The two families also differ in that onions have true bulbs and brodiaeas have corms.

**STATISTICS** Around 300 species widely distributed (mostly) across the northern hemisphere. Besides several species cultivated for their ornamental qualities, several are favorite food and flavoring plants including garlic, onions, shallots, scallions, and chives.

## CALIFORNIA GENERA AND SPECIES

The region has one native and one nonnative genus.

*Allium* (wild onion) has 47 native species and four introduced species. Alliums all have separate tepals. Features to look for in keying out the species include the number and form of the leaves (cylindrical vs. flat); details and presence of crests on the ovaries; and bulb coat patterns examined under a strong hand lens or dissecting microscope. Several of the species are restricted to special geographic areas and habitats but many are widespread and variable.

*Allium validum* (swamp onion) grows to 3 feet tall, has tight umbels of pink-purple flowers, and grows in high mountain meadows.

*A. dichlamydeum* (coast onion) lives on coastal bluffs, has short stems, and tight clusters of narrow, rose-purple flowers.

*A. amplectens* (paper onion) lives on rocky outcrops and barren grasslands, has short stems, and open clusters of white, star-shaped flowers.

*A. haematochiton* (red-skinned onion) lives in grassy areas of southern California, has short stems, rhizomatous bulbs, and pink flowers.

*A. crispum* (crinkled onion; see fig.) lives in oak and other woodlands, has short stems, and open umbels of rose-purple flowers with the inner tepals narrow and crinkled.

*A. triquetrum* ("wild" onion) lives in coastal forests, has tall, three-sided stems, and nodding clusters of bell-shaped white flowers. It is a common escape from gardens.

*Ipheion uniflorum* is a garden escape from South America and features flat leaves and white to pale purple flowers with the tepals joined to form a short tube.

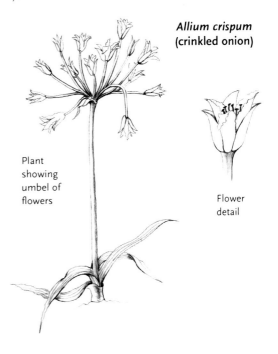

***Allium crispum* (crinkled onion)**

Plant showing umbel of flowers

Flower detail

# ANACARDIACEAE (Sumac Family)

RECOGNITION AT A GLANCE Woody plants with aromatic leaves, dense clusters of tiny, pale flowers, and berrylike drupes.

VEGETATIVE FEATURES Shrubs or trees with simple, leathery or deciduous, compound leaves that are resinously aromatic.

FLOWERS Tiny, white, pinkish, or yellow, often unisexual, and in dense racemes or panicles.

FLOWER PARTS 5 separate sepals, 5 separate petals, (usually) 5 stamens, and a single pistil with a superior ovary on a nectar-secreting disc.

FRUITS Fleshy drupes with a single seed.

RELATED OR SIMILAR-LOOKING FAMILIES There are many families of shrubby plants with small flowers. One possibly confusing family is Rhamnaceae (buckthorn family), which always has simple leaves and often blue or purple flowers, and 3-sided drupes or capsules.

STATISTICS 850 species with a worldwide distribution. Many live in rain forests or seasonal, lowland forests. Several are noted for the toxins and skin irritants in their leaves, including poison-oak and poison-ivy (*Toxicodendron diversilobum* and *T. radicans*). Some are a source of varnish. Mango (*Mangifera indica*), pistachio (*Pistacia vera*), and cashew nut (*Anacardium excelsum*) are widely cultivated food plants. Garden ornamentals include Chinese pistache (*Pistacia chinensis*), Chilean pepper-tree (*Schinus molle*), and smokebush (*Cotinus coggygria*).

## CALIFORNIA GENERA AND SPECIES

The region has three native and two introduced genera.

### NATIVE GENERA

*Toxicodendron diversilobum* (poison-oak; see fig. a) is a woody ground cover, shrub, or vine with toxic, trifoliate leaves, and whitish flowers and fruits. It lives in a wide variety of habitats below 4,000 feet.

*Rhus* (various common names) are woody shrubs with simple or trifoliate, nontoxic leaves; white, pink, or yellow flowers; and sticky, reddish fruits. Species include:

*R. integrifolia* (lemonade berry) has broad, nearly flat, slightly toothed leaves and lives in southern California's coastal sage scrub and chaparral.

*R. ovata* (sugar bush; see fig. b) has broad, folded, toothless leaves and also lives in southern California chaparral.

*R. trilobata* (squawbush, sourberry) is a colonizing shrub with trifoliate, hairy leaves, arching branches, and pale yellow flowers. It is found in canyons and rocky places in the inner foothills and desert mountains.

*Malosma laurina* (laurel sumac) is a tall, woody shrub with simple, folded, evergreen leaves; white flowers; and smooth, dry drupes. It grows in the chaparral of southern California.

### NONNATIVE GENERA

*Schinus molle* (Chilean pepper-tree) is a South American alien with drooping branches, pinnately compound leaves, tiny greenish white flowers, and red drupes. It is widely naturalized in coastal areas.

**a. *Toxicodendron diversilobum* (poison-oak)**

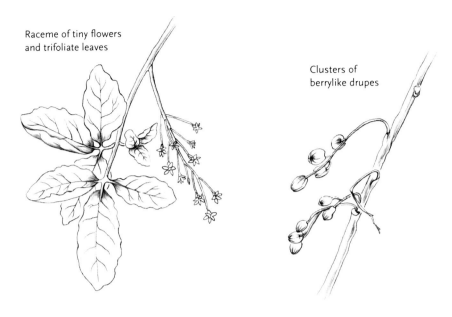

**b. *Rhus ovata* (sugar bush)**

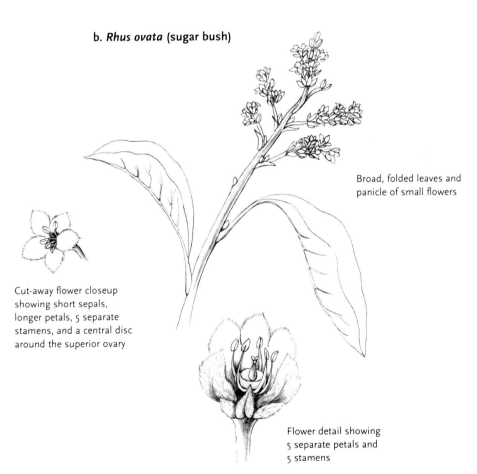

ANACARDIACEAE (SUMAC FAMILY)

# APIACEAE (Parsley or Carrot Family)

Apiaceae was formerly called Umbelliferae.

RECOGNITION AT A GLANCE Highly scented leaves (odor of parsley, dill, anise, and others) with sheathing bases and tiny flowers in compound umbels. Flowers are usually white or yellow.

VEGETATIVE FEATURES Fragrant, herbaceous annuals and perennials with oil tubes in the stems (and sometimes the leaves), and simple to highly compound leaves with a sheathing base.

FLOWERS Tiny, usually bisexual, and arranged in umbels of umbels (compound umbels) or occasionally umbels of heads.

FLOWER PARTS 5 minute sepals, 5 separate petals, 5 stamens, and a single pistil with an inferior, 2-chambered ovary topped with a swollen stylopodium and 2 styles.

FRUITS Schizocarps that split into 2 one-seeded sections.

RELATED OR SIMILAR-LOOKING FAMILIES A few other families bear their flowers in simple, not compound umbels, and are unlikely to be mistaken for this family. Araliaceae (ginseng or aralia family) has umbels of tiny flowers arranged in racemes or panicles, is often woody, and has berrylike fruits.

STATISTICS At least 3,000 species with a worldwide distribution. Several are virulently poisonous including poison hemlock (*Conium maculatum*) and water hemlock (*Cicuta douglasii*). Many are culinary and medicinal herbs such as caraway, dill, fennel, anise, coriander, and parsley. Vegetables include celery, parsnip, and carrot. A few are cultivated for their ornamental foliage or flowers. Several are pernicious and persistent weeds.

## CALIFORNIA GENERA AND SPECIES

The region has 42 genera, several that are nonnative or with nonnative species. The genera are often identified by characteristics of the fruits—the presence and nature of ribs and wings, for example—and are difficult to identify without them. A partial list follows.

*NATIVE GENERA WITH UNUSUAL FEATURES*

*Eryngium* (button-parsley) has prickly leaves, buttonlike heads at the end of an umbel, and tiny green to purple flowers with spiny bracts. It is common in vernal pools and other temporary wetlands.

*Sphenosciadium capitellatum* (ranger's buttons, woolly heads) has large, coarsely pinnately compound leaves, and large umbels ending in white to pale purple heads of flowers. It lives in wet mountain meadows.

*Sanicula* (sanicles) have variable leaves, heads of yellow or red flowers at the end of an umbel, and prickly fruits. Species include:

*S. crassicaulis* (woodland sanicle; see fig. a) has round, palmately lobed and serrated leaves, pale yellow flowers, and lives in wooded places.

*S. bipinnata* (poison sanicle) has coarsely twice divided leaves and yellow flowers, and lives in brushy habitats and open woodlands.

*S. bipinnatifida* (purple sanicle) has bluish green, coarsely almost twice divided leaves, and red-purple (or pale yellow) flowers; it lives in grasslands.

*S. arctopoides* (footsteps-to-spring) grows as a flattened rosette, has yellow-green leaves that are slashed into irregular lobes, and bright yellow flowers in the center of the leaves. It lives on coastal bluffs and in coastal grasslands.

*NATIVE GENERA 4 OR MORE FEET TALL*

*Heracleum lanatum* (cow-parsnip; see fig. b) is a robust perennial to 6 feet high with ragged, pinnately compound leaves and immense compound umbels of white flowers, the outer flowers of each umbellet enlarged. It is widespread in coastal woods and mountain meadows.

*Angelica* (angelicas) are robust perennials with coarse, pinnately compound leaves and large, often globe-shaped compound umbels of white flowers. There are several species from varied habitats.

*Cicuta douglasii* (water hemlock) is a robust perennial with once to twice pinnately compound leaves; large, flat-topped clusters of white flowers; and chambered roots. It is deadly poisonous and lives in marshes and wet meadows.

NATIVE GENERA SHORTER THAN 4 FEET TALL

*Oenanthe sarmentosa* (water parsley) is a colonizing perennial with coarse, pinnately compound leaves, trailing stems, and umbels of white flowers. The fruits turn wine red. It lives in coastal marshes and is poisonous.

*Perideridia* (yampahs) are tuberous-rooted perennials with small, pinnately compound leaves of linear divisions and small, compound umbels of white flowers. The several species are difficult to tell apart; many live in mountain meadows.

*Osmorhiza* (sweet cicely) has three species of small, tough rooted perennials with coarsely divided, anise-scented leaves and open compound umbels of white or pale yellow flowers and bristly, tapered fruits. They live in wooded areas.

*Ligusticum* (lovage) has three species of taprooted perennials with celery-scented, fernlike foliage and white flowers.

*L. apiifolium* is coastal; *L. grayi* lives in mountain meadows.

*Lomatium* (biscuit-root; hog-fennel, and other common names; see fig. c) includes many species of taprooted perennials with parsley or celery scented foliage that varies from highly divided and fernlike to coarsely divided and celerylike. Flowers are mostly yellow (a few have red flowers) and fruits have 2 wings, one on each side. They live in a variety of habitats.

*Tauschia* (tauschia) has three species of taprooted perennials with coarsely divided, celery-scented leaves and rounded leaflets. Open umbels of yellow-green flowers. They live in woodlands.

NONNATIVE GENERA

*Conium maculatum* (poison hemlock) has purple-spotted stems, dissected, ill-scented, fernlike leaves, and white flowers.

*Daucus carota* (Queen Anne's lace; wild carrot) has hairy stems, dissected, carrot-scented, fernlike leaves, and white flowers. The umbels invert in fruit to resemble a basket.

*Foeniculum vulgare* (fennel; see fig. d) has highly dissected, anise-scented leaves with threadlike divisions and yellow flowers.

*Anthriscus caucalis* (bur-chervil) and *Torilis scandicina* (hedge-parsley) are both small annuals with pleasantly scented, fernlike foliage, white flowers, and bristle-covered fruits that catch on clothing.

*Scandix pecten-veneris* (shepherd's needles) is similar in stature and leaf to the last two genera but the outer flowers are irregular, have enlarged petals, and the fruits have needlelike prongs.

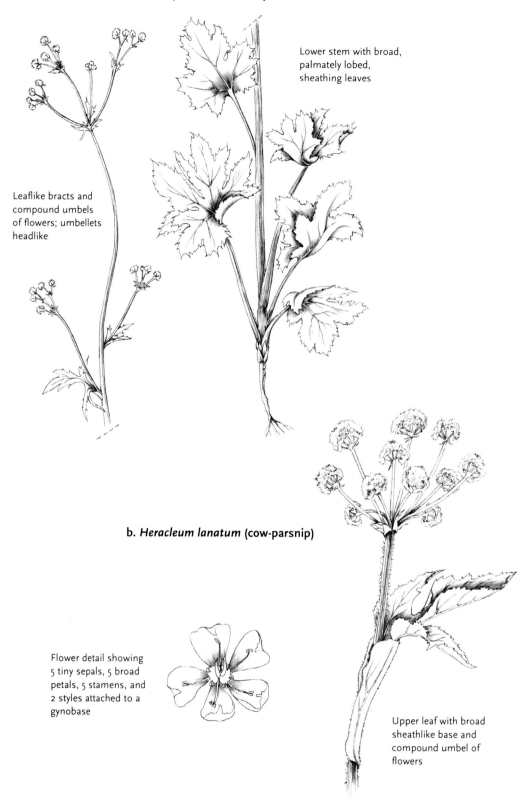

### c. *Lomatium dasycarpum* (hog-fennel)

Compound umbel of fruits, each with 2 lateral wings

### d. *Foeniculum vulgare* (fennel)

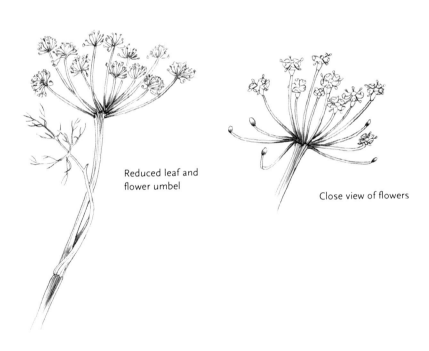

Reduced leaf and flower umbel

Close view of flowers

# APOCYNACEAE (Dogbane Family)

Apocynaceae includes the former Asclepiadaceae, or milkweed family.

RECOGNITION AT A GLANCE Leaves in pairs or whorls with copious milky sap and flowers with a central gynostegium consisting of stamens around or fused to the styles and stigmas.

VEGETATIVE FEATURES Herbaceous perennials with simple, entire, whorled or opposite leaves containing copious poisonous, milky sap.

FLOWERS Symmetrical, variously shaped, curiously constructed, and single or arranged in cymes or umbels. Major pollinators include butterflies, specialized beetles, and bumblebees.

FLOWER PARTS 5 separate upright, spreading, or downturned sepals; 5 spreading or downturned petals; a central gynostegium consisting of 5 stamens and 5 styles, and 2 pistils with superior ovaries. Details of the gynostegium vary; some genera such as the milkweeds have an anther head with 5 embedded anthers containing masses of pollen and 5 enlarged filaments that form hoods (cuplike nectar containers) and the cojoined pistil head has 5 slitlike stigmas between the anthers.

FRUITS 1-chambered follicles with many seeds and attached tufts of long hairs.

RELATED OR SIMILAR-LOOKING FAMILIES The similar milkweeds (formerly Asclepiadaceae) are now considered to belong to this family. The two families were once separated by the complex gynostegium typical of milkweeds, which consists of a combination of modified stamens, styles, and stigmas.

STATISTICS Around 4,000 species, diverse in the tropics and arid parts of southern Asia and Africa. The milkweeds (*Asclepias* spp.) have a broad, mostly temperate distribution. The family is noted for its toxic, cardiac glycosides in stems and leaves; monarch butterfly caterpillars ingest and store compounds from milkweed leaves to defend themselves from being eaten by birds. Many genera are cultivated as ornamentals including periwinkles (*Vinca* and *Catharanthus* spp.), oleander (*Nerium oleander*), and star-jasmine (*Trachelospermum jasminoides*). The succulent stapeliads have bizarre, smelly, starfishlike flowers; *Ceropegia* spp. are unusual vines and succulents with parasol-like flowers; and wax plant (*Hoya* spp.) are succulent, subtropical vines with waxy white or pink flowers. In addition, several genera such as the frangipani (*Plumeria rubra*) are borderline hardy in California gardens. Some of the species are also used medicinally.

## CALIFORNIA GENERA AND SPECIES

The region has seven mostly native genera with 23 species and three nonnative genera and species.

*Apocynum* (dogbane, Indian hemp; see fig. a) has two native species of winter-dormant perennials with fibrous stems, pairs of broadly oval leaves, and clusters of white or pink, bell-shaped flowers. The fibers were widely used for cordage by the California Indians.

*Asclepias* (milkweed) has 14 native species—nonviny perennials with downturned sepals and petals and complex gynostegia.

*A. californica* (California milkweed; see fig. b) has pairs of woolly white leaves on sprawling stems and dark red-purple flowers. It lives on rock outcrops in the arid foothills.

*A. fascicularis* (whorled milkweed) has whorls of narrow leaves and small, pale purple flowers. It is the most widely distributed species.

*A. cordifolia* (heart-leaf milkweed) has pairs of clasping, heart-shaped leaves and nodding dark red and white flowers. It lives in open mixed evergreen and pine forests.

*A. speciosa* (showy milkweed, see fig. c) has pairs of broadly ovate leaves and large, fragrant, pink-purple flowers. It frequents arid roadsides and low-elevation mountain meadows.

*Cycladenia humilis* is a small perennial herb with broad, bluish green leaves and clusters of attractive pink-purple flowers. It is found in rocky habitats in the mountains.

*Vinca major* (periwinkle; see fig. d) is a non-native, Mediterranean ground cover with large, single, blue, cup-shaped flowers. It has invaded natural areas near gardens.

### a. *Apocynum cannabinum* (Indian hemp)

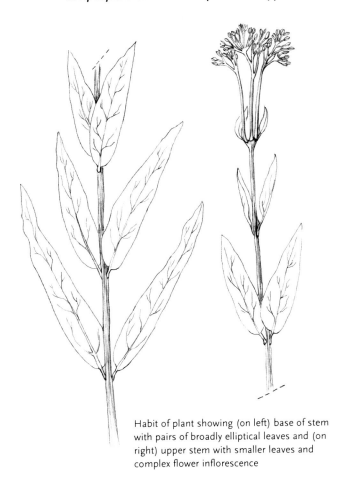

Habit of plant showing (on left) base of stem with pairs of broadly elliptical leaves and (on right) upper stem with smaller leaves and complex flower inflorescence

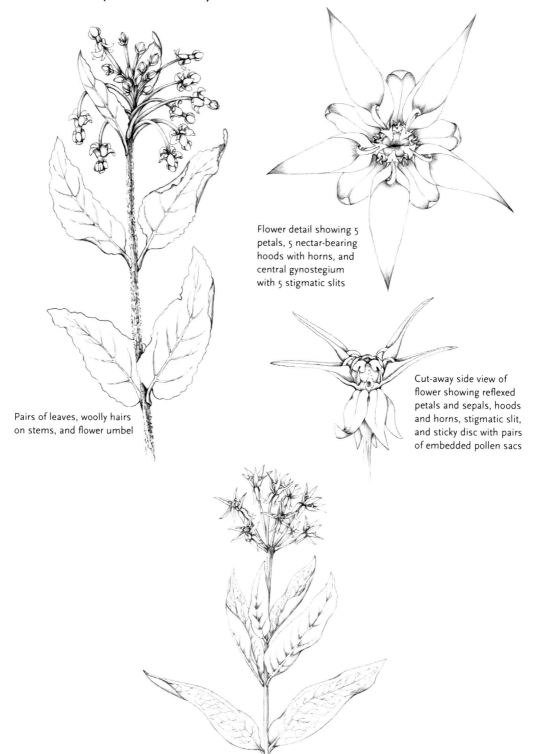

**b. *Asclepias californica* (California milkweed)**

Pairs of leaves, woolly hairs on stems, and flower umbel

**c. *Asclepias speciosa* (showy milkweed)**

Flower detail showing 5 petals, 5 nectar-bearing hoods with horns, and central gynostegium with 5 stigmatic slits

Cut-away side view of flower showing reflexed petals and sepals, hoods and horns, stigmatic slit, and sticky disc with pairs of embedded pollen sacs

Upper stem with pairs of broad leaves and flower umbel

**d. *Vinca major* (periwinkle)**

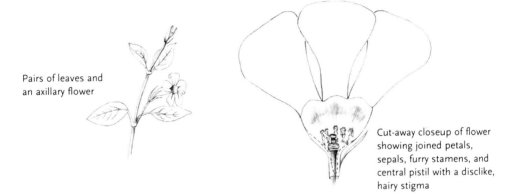

Pairs of leaves and an axillary flower

Cut-away closeup of flower showing joined petals, sepals, furry stamens, and central pistil with a disclike, hairy stigma

# ARECACEAE (Palm Family)

Arecaceae was formerly called Palmae.

RECOGNITION AT A GLANCE Mostly single-trunked trees with top crowns of enormous, palmately lobed to pinnately compound leaves.

VEGETATIVE FEATURES Multi-trunked shrubs or single-trunked trees with terminal clusters of enormous, palmately lobed or compound to feathery, pinnately compound leaves. Trunks never increase in girth but remain the same diameter throughout their lives. The leaves often leave behind characteristic leaf bases and/or scars when they fall.

FLOWERS Small, numerous, mostly white or cream-colored, and borne in panicles that emerge from large, boat-shaped bracts below or among the crown of leaves.

FLOWER PARTS 6 tepals, (usually) 6 stamens, and a single pistil with a superior (sometimes inferior) ovary.

FRUITS Fleshy, 1-seeded drupes that vary in size from small and datelike to huge and similar to coconuts.

RELATED OR SIMILAR-LOOKING FAMILIES The palm family is unique in appearance, and combines a single, uniform trunk with a mop of enormous, divided leaves. Occasionally, members of the genera *Dracaena* and *Cordyline* are confused with palms but their leaves are never divided or compound.

STATISTICS 3,000 species distributed throughout the world's tropics; a few species also occur in warm temperate areas. A very important family in the tropics for its many edible fruits—particularly the coconut and date palms—and for seeds rich in oil. In the tropics, palms also serve a multitude of other purposes from medicines to thatching, starches to needles, rattans to sugar.

## CALIFORNIA GENERA AND SPECIES

The region has one native genus and species, and two nonnative genera with three species that are sometimes naturalized.

*Washingtonia filifera* (California fan palm, see fig.) lives in oases at the base of rugged desert mountains. Look for it in Joshua Tree National Park, Anza Borrego Desert State Park, and the Palm Springs area. The small, datelike fruits were eaten by the Indians and their pits were ground and roasted into flour.

*Washingtonia robusta* (Mexican fan palm) is closely related to our native fan palm but is native to Arizona and northern Mexico. It is commonly planted along southern California boulevards. Its trunks are taller and more slender than our native fan palm.

*Phoenix canariensis* (Canary Island date palm) and *P. dactylifera* (African date palm) are also reported as occasionally naturalized. Both have pinnately compound leaves and are widely cultivated, *P. canariensis* as an ornamental and *P. dactylifera* for its fruits.

## *Washingtonia filifera* (California fan palm)

Tree habit with mop of deeply palmately lobed leaves and old leaf thatch

Leaf detail showing spiny petiole and fanlike divisions

# ASTERACEAE (Daisy, Composite, or Sunflower Family)

RECOGNITION AT A GLANCE Heads of flowers that resemble a single larger flower and the (usual) presence of a pappus on the tiny, seed-like fruits.

VEGETATIVE FEATURES Annuals, perennial herbs, and small shrubs with a wide variety of leaf designs and arrangements. Many have leaves containing volatile, fragrant oils.

FLOWERS Arranged in tight heads to resemble single, large flowers. There are three basic designs: heads with strap-shaped ray flowers only (example, dandelion), heads with tiny star-like disc flowers only (example, ageratum), and heads with peripheral ray flowers around central disc flowers (examples, marguerites and daisies). Heads are surrounded by one or more rows of usually green, sepal-like bracts called phyllaries.

FLOWER PARTS Sepals are modified into hairs, bristles, or scales (*pappus*) or are missing; ray flowers have 5 highly fused, flat, strap-shaped petals; disc flowers have 5 starlike petals joined to form a tube. There are 5 stamens usually fused by their anthers to form a hollow tube, and a single pistil with an inferior ovary and 2 style branches. Good magnification is required to see these features. Often, ray flowers are sterile and disc flowers are fertile, but there are many exceptions.

FRUITS Tiny, 1-seeded achenes that resemble seeds. Most achenes have a pappus attached to the top for wind dispersal.

RELATED OR SIMILAR-LOOKING FAMILIES This distinctive family is usually readily recognized and seldom confused with others. A few other families have evolved similar strategies of massing many flowers together for show but have different flower details. Among them, some of the dogwoods (*Cornus* spp.) feature a buttonlike head of 4-petalled flowers surrounded by petal-like white or pink bracts. Yerba mansa (*Anemopsis californica* in the Saururaceae) features a fat spike of petalless flowers surrounded at the base by several white, petal-like bracts.

STATISTICS Immense family of possibly 20,000 species distributed throughout the world and dominant in most floristic provinces and floras except for tropical rain forests. Many aggressive and weedy species belong to the family and are aided by the efficient wind dispersal of their tiny, 1-seeded fruits. A few edible species include chicory, radicchio, and endive (*Cichorium* spp.), lettuce (*Lactuca sativa*), globe artichoke (*Cynara cardunculus*), Jerusalem artichoke (*Helianthus tuberosus*), and sunflower seeds and oil (*Helianthus annuus*). Many popular garden flowers belong to numerous genera and species, including aster (*Callistephus* and *Aster* spp.), daisies (*Erigeron* spp.), marigolds (*Tagetes* spp.), zinnias (*Zinnia* spp.), cosmos, dahlias, marguerites, chrysanthemums, and many others. Culinary and medicinal herbs include tansy (*Tanacetum vulgare*), tarragon (*Artemisia dracunculus*), wormwood (*A. absinthium*), and yarrow (*Achillea millefolium*).

## CALIFORNIA GENERA AND SPECIES

Owing to its size, most books split the family into tribes, then genera and species. With new studies, the tribes are being reinterpreted and many are no longer considered natural, evolutionary groups. The most prominent tribes follow.

### ANTHEMIDEAE (MAYWEED TRIBE)

Leaves usually highly aromatic, often deeply divided or compound, flowers with or without rays, pappus missing or highly reduced.

*Achillea millefolium* (yarrow) has fernlike leaves and flat-topped clusters of tiny white flower heads.

*Artemisia* (sagebrush) has variable, sage-scented leaves and spikes of tiny, wind-pollinated disc flowers.

*Chamomilla suaveolens* (pineapple weed) [now *Matricaria discoidea*] has highly dissected bright green leaves and thimble-shaped heads of disc flowers.

*Anthemis cotula* (mayweed) has similar but ill-scented leaves and white ray flowers around the cone-shaped clusters of disc flowers.

*ASTERFAE (ASTER TRIBE)*
Large tribe. Leaves highly variable; flowers with or without rays; phyllaries in two to several, usually overlapping, shinglelike rows; flowers have a (usually) hairy pappus.

PROMINENT GENERA WITH MOSTLY YELLOW FLOWERS *Solidago* (goldenrod) has panicles of tiny yellow flowers.

*Ericameria* (goldenbushes) are small, evergreen shrubs with yellow, daisylike flower heads.

*Heterotheca* (golden asters) are (mostly) mounded perennials with yellow flowers.

*Grindelia* (gumplants) are perennials with large yellow daisies whose buds are covered with a white gum.

*Hazardia* and *Isocoma* (also known as goldenbushes) are two more small, evergreen shrubs with yellow daisies.

PROMINENT GENERA WITH USUALLY WHITE, PINK, PURPLE, OR BLUE RAYS *Aster* (asters) and *Erigeron* (daisies or fleabanes; see fig. a) are varied perennials with similar-looking heads of yellow disc flowers and (usually) white, blue, or purple ray flowers. Asters generally have fewer, coarser rays.

*Bellis perennis* (English daisy) is a naturalized, coastal, tufted perennial with white rays and yellow discs.

*Lessingia* (various common names) are annuals and perennials of varied appearance. *L. filaginifolia* (woolly aster) has grayish, woolly leaves and pink-purple rays around yellow discs.

*Baccharis* (coyote bush, mulefat, and others) is unusual in being dioecious; its discoid flower heads are white to pale pink.

*CICHORIEAE (CHICORY TRIBE)*
Another large tribe. The combination of only ray flowers and leaves with milky sap are distinctive for the family. The details of leaves, phyllaries, and pappus vary widely.

NATIVE GENERA *Stephanomeria* (wand flower) includes annuals and perennials with pinkish flowers.

*Hieracium* (hawkweeds) are perennials with white or yellow flowers and furry leaves.

*Microseris* and *Agoseris* (native dandelions; see fig. b) are annuals and perennials with yellow flower heads that grow in natural rather than disturbed habitats.

*Uropappus lindleyi* (silver puffs) is an annual with pale yellow flowers and a pappus of narrow, silvery, pointed scales.

NUMEROUS INTRODUCED WEEDS *Taraxacum officinale* (common dandelion) has a rosette of hairless basal leaves and single yellow flower heads. *Hypochaeris* (cat's ear; see fig. c) looks similar to dandelion but has hairy leaves and branched stalks bearing flower heads.

*Sonchus* (sow thistles) are leafy annuals with toothed, sometimes prickly leaves and small, pale yellow flower heads.

*Picris echioides* (prickly ox-tongue) is a much-branched annual with bristly haired leaves and small yellow flower heads.

*Cichorium intybus* (chicory) is a much-branched perennial with smooth leaves and clear blue flowers.

*CYNAREAE (THISTLE TRIBE)*
Mostly simple leaves that are often heavily armed; (usually) spine-tipped phyllaries; long, tubular disc flowers only; and a variable pappus.

INTRODUCED AND AGGRESSIVE WEEDS *Silybum marianum* (milk-thistle) is a robust plant with spiny leaves marbled with white veins and purple flowers.

*Centaurea* (star-thistles) are aggressive annuals with nonspiny leaves and spiny phyllaries around purple or yellow flowers.

*Cirsium vulgare* (bull thistle; see fig. d) is a robust biennial with spiny leaves and showy, red-purple flower heads surrounded by spiny phyllaries.

*Cynara cardunculus* (cardoon) resembles an artichoke, with bold, grayish, spiny leaves and very stout flowering stalks with very large blue-purple flowers.

*Carduus* (plumeless thistles) are annuals with spiny stems and leaves and small heads of purple flowers with spiny phyllaries.

*Cirsium* (true thistles) has several native species including the widespread, variable cobweb thistle (*C. occidentale;* see fig. e), which has pink, rose-purple, or bright red flowers and cobwebby hairs on phyllaries and leaves.

### HELIANTHEAE (SUNFLOWER TRIBE)

Variable leaves, one or few rows of phyllaries, flowers with or without rays, and a pappus of few scales or missing. Unlike most other tribes, there are often internal bracts between the disc flowers. Bracts are directly attached to the receptacle and should not be confused with the pappus, which is always on top of the ovary.

*Helianthus* (sunflower) are annuals or perennials with rough-textured stem leaves and large heads of yellow rays and often dark purple or red disc flowers.

*Helianthella* (little sunflower) are perennials with mostly basal leaves and large heads with yellow ray flowers carried just above the foliage.

*Wyethia* (mule's ears; see fig. f) are rhizomatous perennials with dramatic, usually broad basal leaves, and very large heads with yellow ray flowers.

*Balsamorhiza* (balsamroot) is similar to mule's ears but has arrowhead-shaped leaves.

*Coreopsis* are perennials or annuals with feathery, fernlike foliage and heads with yellow ray flowers; the phyllaries have two different shapes.

*Rudbeckia californica* (coneflower) is a meadow perennial with arrowhead-shaped leaves, tall stalks, and yellow rays around an elongated cone of dark disc flowers.

**AMBROSIINAE (RAGWEED SUBTRIBE)** Unisexual, usually petalless, wind-pollinated disc flowers only.

*Dicoria* (poverty weed) is a weedy native of sandy and salty soils with male and female flowers mixed in the same head. (Other genera have male and female flowers in separate heads).

*Xanthium* (cocklebur) has weedy annuals that grow in disturbed, temporarily wet places; simple leaves, and spiny, burlike female flowers.

*Ambrosia* (ragweed) has annuals, perennials, and small shrubs with (usually) deeply lobed leaves and spiny, burlike female flowers.

*A. chamissonis* (dune bursage) is a mat-forming, gray, woody perennial found on coastal sand dunes.

**MADIINAE (TARPLANTS SUBTRIBE)** Foliage and phyllaries usually have fragrant, glandular hairs. The phyllaries wrap part or all the way around the ovary of each ray flower.

*Achyrachaena mollis* (blow wives) is a spring annual with orange flowers and a conspicuous, flowerlike fruiting head with two unequal rows of pearly white pappus scales.

*Layia* (tidy tips) are spring annuals with conspicuous white, yellow, or yellow-tipped white ray flowers.

*Madia* (see fig. g), *Hemizonia* [now split also into *Deinandra*], *Holocarpha*, and others (tarplants) are mostly late-spring and summer annuals with a pungent fragrance, very sticky glands, and heads with yellow or white rays. They are difficult to differentiate without close inspection of the phyllaries and pappus.

*Calycadenia* (rosinweed) [now mostly in *Osmadenia*] are summer annuals with conspicuous glands and ray flowers that are split into 3 fanlike segments.

**HELENIEAE (SNEEZEWEED SUBTRIBE)** Variable leaves, one or few rows of phyllaries, flowers with or without rays, and a pappus crown of scales (sometimes missing).

*Lasthenia* (goldfields) are spring annuals with opposite leaves and bright yellow ray flowers.

*Blennosperma nanum* (glueseed) is a spring annual with pale yellow rays and white pollen.

*Eriophyllum* (golden-yarrow, lizard tail, and others; see fig. h) are spring and summer annuals and woody perennials with woolly hairs on usually lobed or dissected leaves and yellow flowers.

*Monolopia* are spring annuals with large yellow rays.

*Helenium* (sneezeweed; see fig. i) are wet-growing perennials with decurrent leaves and dome-shaped heads of disc flowers surrounded by yellow rays.

*Chaenactis* (pincushion flowers; see fig. j) are spring annuals or summer perennials with deeply lobed leaves and (usually) white to pink disc flowers.

### INULEAE (EVERLASTING TRIBE)

Simple leaves densely clothed (usually) in tangled, woolly, white hairs; phyllaries in several rows and either dry and papery or covered in dense white wool; discoid heads only (no ray flowers), and hairy pappus.

*Anaphalis margaritacea* (pearly everlasting; see fig. k) is a rhizomatous, colonizing summer perennial with dense clusters of flowers with pearly white phyllaries.

*Gnaphalium* (cudweed) are introduced and native annuals and taprooted perennials with strongly scented foliage and flower heads similar to pearly everlasting or with phyllaries covered in dense wool.

*Antennaria* (pussy toes) are matted perennials from mountain meadows with basal leaves and flat-topped clusters of tiny, flowers like pearly everlasting or *Anaphalis*.

Many "belly" annuals in several genera are confusing to sort out. They occur in hard-packed soils and temporarily moist places throughout the foothills.

### SENECIONEAE (SENECIO TRIBE)

Variable leaves, flower heads with one even row of nonoverlapping phyllaries (additional tiny phyllaries sometimes are at the base of the heads), with or without ray flowers, and a hairy pappus.

*Senecio* (groundsel, butterwort, and others; see figs. l and m) is a large genus typified by alternate (or basal) leaves of several designs and heads with or without ray flowers. Species live in many different habitats throughout the state.

*Arnica* (arnica) are perennials with pairs of usually broad, simple leaves and flower heads with or without rays. Most live in the mountains. (Some botanists are now placing *Arnica* in the tarweed subtribe, *Madiinae*.)

*Raillardella* are matted woody perennials with simple, alternate leaves, and narrow heads of disc flowers.

*Petasites frigidus* var. *palmatus* (western coltsfoot) is a rhizomatous perennial with large round, deeply palmately lobed leaves and clusters of white to pale pink flower heads with tiny rays.

*Adenocaulon bicolor* (trail plant) is a clump-forming perennial with triangular leaves green above and silvery underneath. Open panicles of minute disc flowers appear in summer.

## ASTEREAE (ASTER TRIBE)
### a. *Erigeron glaucus* (seaside daisy)

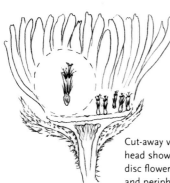

Cut-away view of flower head showing hairy bracts, disc flowers on receptacle, and peripheral ray flowers

Detail of disc flower showing inferior ovary, hairy pappus, tubular petals, and 2-forked style

Ray flower

## CICHORIAEA (CHICORY TRIBE)
### b. *Agoseris* sp. (native dandelion)

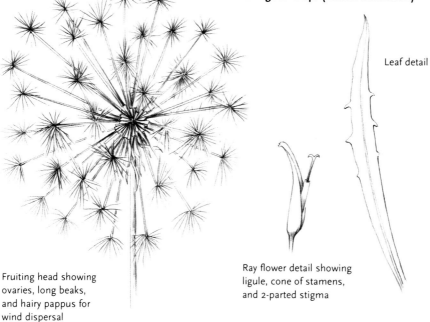

Fruiting head showing ovaries, long beaks, and hairy pappus for wind dispersal

Ray flower detail showing ligule, cone of stamens, and 2-parted stigma

Leaf detail

Flower head showing rows of bracts and numerous ray flowers

### c. *Hypochaeris radicata* (hairy cat's ear)

Ligulate flower head showing all ray flowers

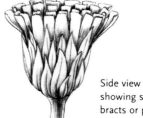

Side view of flower head showing several rows of bracts or phyllaries

## CYNAREAE (THISTLE TRIBE)

### d. *Cirsium vulgare* (bull thistle)

### e. *Cirsium occidentale* (cobweb thistle)

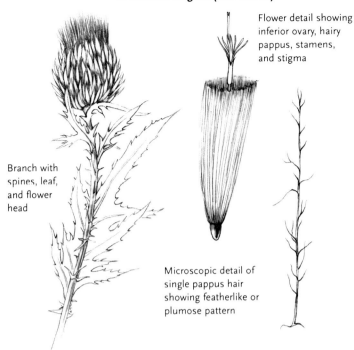
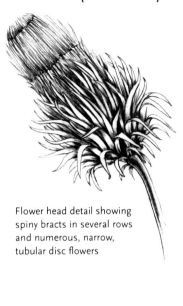

Branch with spines, leaf, and flower head

Flower detail showing inferior ovary, hairy pappus, stamens, and stigma

Microscopic detail of single pappus hair showing featherlike or plumose pattern

Flower head detail showing spiny bracts in several rows and numerous, narrow, tubular disc flowers

## HELIANTHEAE (SUNFLOWER TRIBE)

### f. *Wyethia helenioides* (woolly mule's ear)

### HELIANTHEAE SUBTRIBE MADIINAE
(Tarweed subtribe of sunflower tribe)

### g. *Madia elegans* (elegant tarweed)

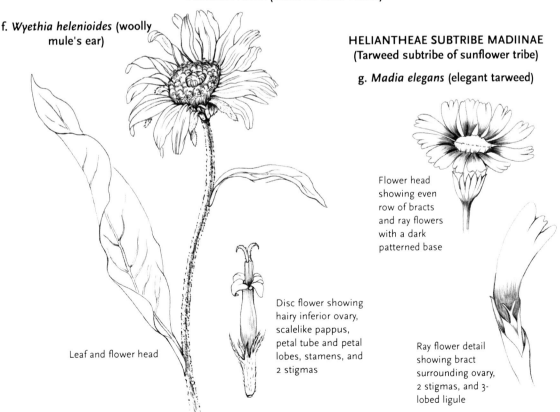

Leaf and flower head

Disc flower showing hairy inferior ovary, scalelike pappus, petal tube and petal lobes, stamens, and 2 stigmas

Flower head showing even row of bracts and ray flowers with a dark patterned base

Ray flower detail showing bract surrounding ovary, 2 stigmas, and 3-lobed ligule

ASTERACEAE (DAISY, COMPOSITE, OR SUNFLOWER FAMILY)

## HELIANTHEAE SUBTRIBE HELENIAE
(Sneezeweed subtribe of sunflower tribe)

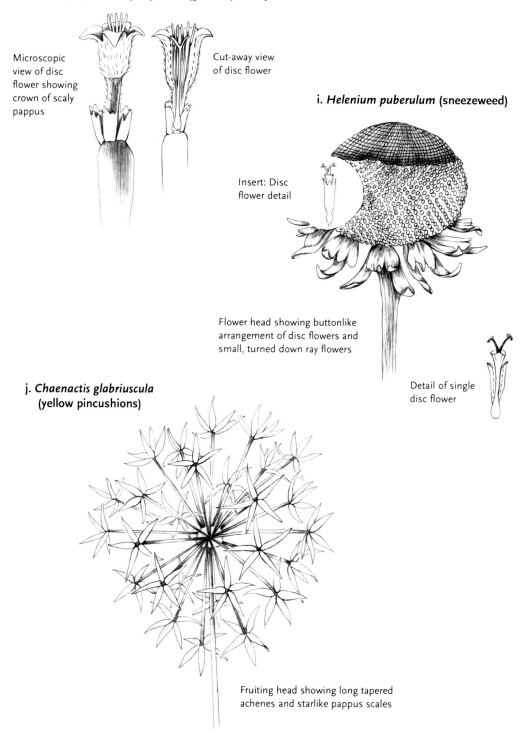

h. *Eriophyllum confertiflorum* (golden yarrow)

Microscopic view of disc flower showing crown of scaly pappus

Cut-away view of disc flower

i. *Helenium puberulum* (sneezeweed)

Insert: Disc flower detail

Flower head showing buttonlike arrangement of disc flowers and small, turned down ray flowers

Detail of single disc flower

j. *Chaenactis glabriuscula* (yellow pincushions)

Fruiting head showing long tapered achenes and starlike pappus scales

## INULEAE (EVERLASTING TRIBE)

### k. *Anaphalis margaritacea* (pearly everlasting)

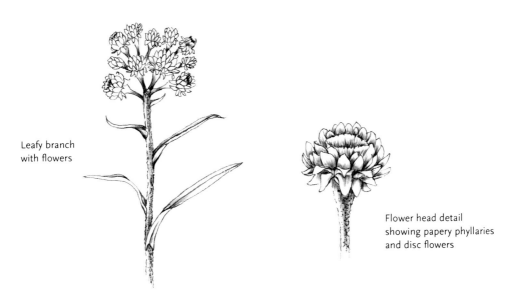

Leafy branch with flowers

Flower head detail showing papery phyllaries and disc flowers

## SENECIONEAE (SENECIO TRIBE)

### l. *Senecio flaccidus* var. *douglasii* (bush senecio)

Flower head showing even row of phyllaries, and ray and disc flowers

### m. *Senecio vulgaris* (common groundsel)

Achene with hairy pappus

# BERBERIDACEAE (Barberry Family)

**RECOGNITION AT A GLANCE** Shrubs with pinnately compound, hollylike leaflets or creeping ground covers; flowers with several rows of colored tepals in threes.

**VEGETATIVE FEATURES** Woody ground covers and evergreen shrubs with creeping rootstocks and compound leaves. Leaves may be in threes or pinnately compound, and often have spiny teeth.

**FLOWERS** Small, white or yellow, and in racemes or complex panicles.

**FLOWER PARTS** 3 or more rows of similar-looking tepals in threes (no distinction between sepals and petals), stamens in threes or multiples of 3, and a single pistil with a superior ovary.

**FRUITS** Bowling pin–shaped berries or capsules; seeds sometimes have attached oil bodies.

**RELATED OR SIMILAR-LOOKING FAMILIES** The barberry family is a primitive and unusual dicot family easily distinguished from others by the unusual floral formula. Although it shares flower parts in threes with the birthwort family (Aristolochiaceae), the leaf design and flower shapes are entirely different. (The Aristolochiaceae is not detailed in this book.)

**STATISTICS** 650 species, mostly in the temperate regions of the world, with great diversity in South America, East Asia, and eastern North America. The berries of some *Berberis* species are edible but the main use of the family is as garden ornamentals, including many species of *Berberis* (barberry, including *Mahonia*), heavenly bamboo (*Nandina domestica*), and *Epimedium* from eastern Asia.

## CALIFORNIA GENERA AND SPECIES

The region has three native genera.

*Berberis* (barberry, aka *Mahonia*; see fig. a) are shrubs with pinnately compound leaves, hollylike leaflets, and yellow flowers. The several species range from coastal bluffs to conifer forests, chaparral, and even deserts.

*Achlys* (vanilla leaf, sweet-after-death, two species) is a creeping, rhizomatous ground cover with trifoliate leaves, leaflets shaped like butterfly wings, and narrow spikes of tiny, white, petalless flowers. It lives in conifer forests in northwestern California.

*Vancouveria* (redwood-ivy, inside-out flower, three species; see fig. b) are rhizomatous ground covers with twice pinnately compound leaves, rounded leaflets, and open panicles of white or yellow flowers with the perianth swept backwards. They thrive in coniferous forests in northern California and the Klamath Mountains.

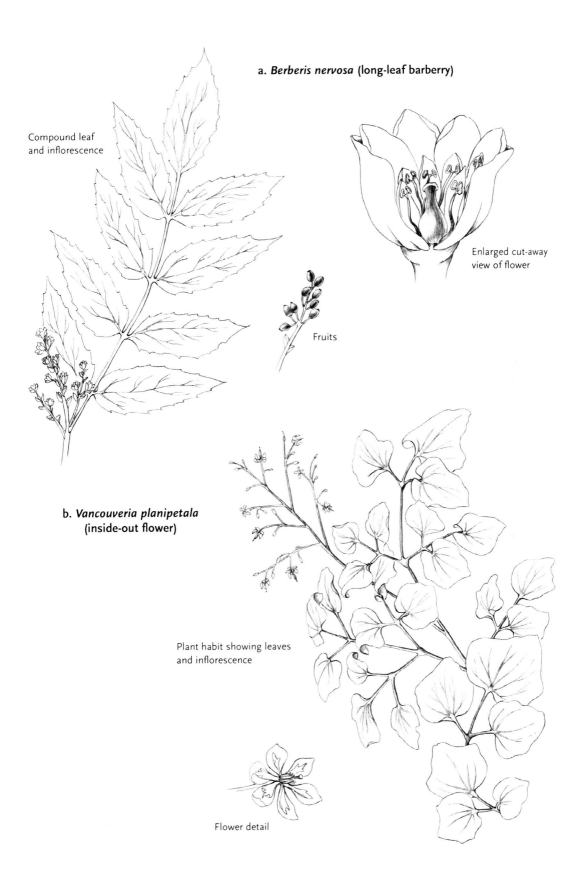

# BETULACEAE (Birch Family)

**RECOGNITION AT A GLANCE** Monoecious shrubs or trees with ovate, coarsely toothed leaves; and petalless, green male flowers in slender, hanging, catkins.

**VEGETATIVE FEATURES** Fast-growing, monoecious, deciduous shrubs and trees from moist habitats. The leaves are simple, ovate, and often doubly serrate (with coarse and fine teeth).

**FLOWERS** Minute, unisexual, greenish, wind-pollinated, and borne in conelike catkins (sometimes also singly or in pairs).

**FLOWER PARTS** Male flowers have bracts and bunches of stamens; female flowers have bracts and a single pistil with feathery or sticky stigmas.

**FRUITS** Female catkins ripen in fruiting spikes that resemble small cones that consist of scalelike bracts and tiny, winged achenes (resemble seeds). Those without female catkins ripen into nuts enclosed in fuzzy bracts.

**RELATED OR SIMILAR-LOOKING FAMILIES** Other families with pendulous male catkins include Garryaceae (garrya family), Salicaceae (willow family), and Fagaceae (oak family). Garryas have tough, opposite, evergreen leaves; the dioecious cottonwoods and willows have seeds covered with white hairs; and the oaks have tough, evergreen leaves or deciduous, lobed leaves, and bear acorns in scaly cups.

**STATISTICS** 120 species across the northern hemisphere in riparian habitats and temperate, broadleaf forests. Filberts and hazelnuts (*Corylus* spp.) are grown for their richly flavored nuts. Birches (*Betula* spp.), alders (*Alnus* spp.), and hornbeams (*Carpinus* spp.) are widely grown in gardens.

## CALIFORNIA GENERA AND SPECIES

The region has three native genera and seven species.

*Corylus cornuta* var. *californica* (California hazelnut; see fig. a) is a small, multi-trunked tree with soft foliage and edible nuts. It lives in moist conifer forests.

*Betula* (birch) is a small, multi-trunked tree or large shrub with firmer leaves and female catkins that shatter when ripe.

*B. occidentalis* (water birch) is common along streams on the east side of the southern Sierra and sporadic elsewhere.

*B. glandulosa* (resin birch) is an uncommon shrub, found in wet meadows in the northeastern corner of the state.

*Alnus* (alders) are shrubs and trees with female catkins that resemble redwood seed cones, and fall intact.

*A. rubra* (red alder) and *A. rhombifolia* (white alder) are riparian trees with a similar appearance; red alder typifies coastal canyons; white alder lives inland along watercourses.

*A. incana* var. *tenuifolia* (mountain alder; see fig. b) is a shrub that forms thickets along streams in the mountains.

### a. *Corylus cornuta* var. *californica* (California hazelnut)

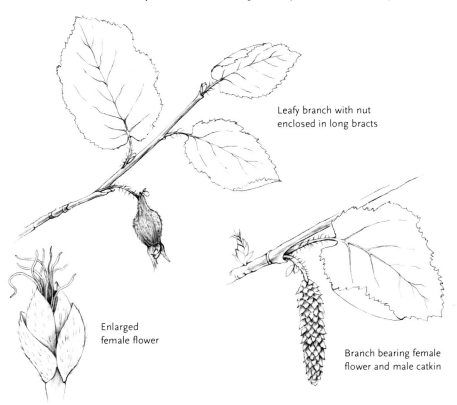

Leafy branch with nut enclosed in long bracts

Enlarged female flower

Branch bearing female flower and male catkin

### b. *Alnus incana* var. *tenuifolia* (mountain alder)

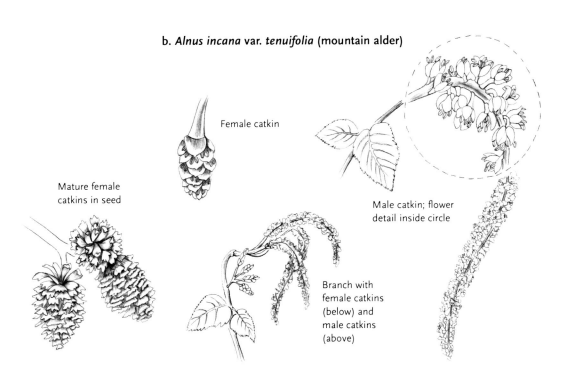

Female catkin

Mature female catkins in seed

Male catkin; flower detail inside circle

Branch with female catkins (below) and male catkins (above)

BETULACEAE (BIRCH FAMILY)

# BORAGINACEAE (Borage or Forget-Me-Not Family)

RECOGNITION AT A GLANCE Bristly-haired plants with "fiddlehead" coils of flower buds, a single style, and a 4-lobed ovary.

VEGETATIVE FEATURES Herbaceous annuals and perennials with simple, alternate leaves often with bristly hairs.

FLOWERS Small, open-bell-shaped, and arranged in tightly coiled, scorpioid cymes that unroll as flower buds open.

FLOWER PARTS 5 partly fused sepals, 5 petals fused to form a tube, 5 stamens attached to the tube, and a single pistil with a superior ovary of 4 segments.

FRUITS 4 one-seeded nutlets (some may abort).

RELATED OR SIMILAR-LOOKING FAMILIES Borages are closely related to the waterleaf family (Hydrophyllaceae), which should be included in the borage family according to recent studies. The major distinction between the families is that borages generally have a single style and a 4-lobed ovary, each lobe containing a single seed. Waterleafs have a 2-forked style and an unlobed ovary containing several to numerous seeds.

STATISTICS 2,000 species with a worldwide distribution; especially diverse in the Mediterranean region and the western United States. Among the many growth forms, the family has tropical trees (genus *Cordia*) with durable hardwood, and several garden ornamentals including heliotrope (*Heliotropium* spp.), borage (*Borago officinalis*), forget-me-not (*Myosotis* spp.), blue-eyed Mary (*Omphalodes*), tower-of-jewels and Pride-of-Madeira (*Echium* spp.), and lungwort (*Pulmonaria* spp.). Several have been used medicinally, including comfrey (*Symphytum officinalis*).

## CALIFORNIA GENERA AND SPECIES

The region has 18 genera and many species; many are difficult to key and require clear magnification of the fruiting nutlets and their attachment.

*WHITE-FLOWERED GENERA*

*Cryptantha* and *Plagiobothrys* (popcorn flowers) are mostly annuals with many species. The two genera require detailed examination of the nutlets to distinguish. Generally cryptanthas live in dry places and plagiobothryses in temporarily wet soils. Plagiobothryses often stain purple. A few cryptanthas are clump-forming perennials in dry mountains.

*Pectocarya* (comb-fruits) look superficially like cryptanthas but have nutlets that fan out and are lined with comblike prickles.

*Heliotropium curassavicum* (heliotrope) is a fleshy-leaved annual or perennial and has flowers with a yellow eye that ages purple. It lives in sandy places.

*BLUE-, PINK-, OR PURPLE-FLOWERED GENERA*

*Cynoglossum* (hound's tongue), perennials with tongue-shaped basal leaves and blue, forget-me-not-like flowers. They live in woods.

*Myosotis* (forget-me-nots), classical garden perennials that escape into moist woods, with oval leaves and sky-blue flowers with yellow centers.

*Mertensia* (mountain bluebells; see fig. a), clump-forming perennials from mountain meadows, with elliptical leaves and nodding, bell-shaped, blue flowers.

*YELLOW- OR ORANGE-FLOWERED GENERA*

*Amsinckia* (fiddlenecks; see fig. b), annuals from the foothills with yellow-orange or orange flowers.

*Lithospermum* (puccoon), perennials from dry rocky mountains, with clear yellow flowers.

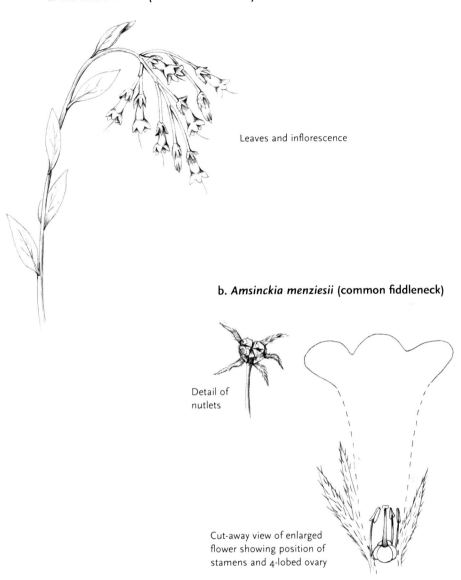

a. *Mertensia ciliata* (mountain bluebells)

Leaves and inflorescence

b. *Amsinckia menziesii* (common fiddleneck)

Detail of nutlets

Cut-away view of enlarged flower showing position of stamens and 4-lobed ovary

# BRASSICACEAE (Mustard Family)

Brassicacaeae was formerly called Cruciferae.

RECOGNITION AT A GLANCE Herbaceous plants with 4 petals arranged like a cross, 6 stamens of 2 lengths, and a long slender silique or rounded, short silicle-type fruits.

VEGETATIVE FEATURES Herbaceous plants (occasionally small shrubs) with simple to pinnately compound leaves often with a peppery flavor. Leaf hairs may be simple or forked.

FLOWERS Crosslike and arranged in racemes or panicles, and are often open at the same time fruits are developing below.

FLOWER PARTS 4 separate sepals and 4 crosslike, clawed petals; 6 stamens usually of 2 lengths; and a single pistil with a superior ovary separated lengthwise by a papery partition.

FRUITS Long, slender siliques or round to squat silicles containing many seeds in 2 or 4 rows.

RELATED OR SIMILAR-LOOKING FAMILIES Mustards are closely related to the caper family (Capparaceae), which is now often included as part of the mustard family. Capers differ in usually having palmately compound or divided, smelly leaves, and a long-stalked gynophore on the flower that carries the ovary well beyond the petals. Several other families feature flowers with 4 petals but seldom in a crosslike arrangement and with 6 stamens of 2 lengths.

STATISTICS Over 3,000 species with a worldwide distribution, most abundant in rocky places and mountains in the northern hemisphere. Diverse in the Mediterranean region and Middle East. The family is of economic importance for mustard oil and seed (*Brassica* spp.) and the vegetable crops broccoli, cauliflower, brussels sprouts, kale, cabbage, and kolhrabi (all forms of *B. oleracea*) as well as rutabagas and turnips. Several common garden ornamentals include wallflowers (*Cheiranthus* spp.), stock (*Matthiola* spp.), and golden basketflower (*Aurinia saxatilis*).

## CALIFORNIA GENERA AND SPECIES

The region has 63 genera with many species. Several are nonnative and weedy or garden escapes. Keys rely heavily on details of the fruits. A few representative genera follow.

*DISTINCTIVE GENERA*

*Streptanthus* (jewelflower; see fig. a) are annuals and perennials with racemes of red, white, pink, or purple flowers with inflated sepals, crimped or twisted petals, and flattened seed pods. They often live on volcanic and serpentine rocks.

*Caulanthus* (also known as jewelflowers) are annuals with similar flowers but have cylinder-shaped seed pods.

*TYPICAL GENERA WITH SLENDER SILIQUE-TYPE SEED PODS*

*Brassica* (mustards; nonnative; see fig. b) are annual weeds with racemes of bright yellow flowers. They were introduced as cover crops and for their oil-rich seeds.

*Raphanus* (wild radish) are annual weeds with racemes of purple, whitish, pink, or cream-colored flowers. They were introduced as a garden vegetable and "escaped" into disturbed habitats.

*Barbarea orthoceras* (winter cress) is a native annual with flowers similar to mustard but with pinnately compound leaves whose terminal leaf is greatly enlarged.

*Rorippa nasturtium-aquaticum* (watercress) is a native perennial that lives in sluggish streams; it features pinnately compound, peppery-flavored leaves and white flowers.

*Cardamine californica* (milkmaids) is a native, tuberous perennial in forests and woodlands, with racemes of early-spring white to purple flowers.

*Erysimum* (wallflowers; see fig. c) are native biennials from dunes, coastal bluffs, and woodlands with racemes of fragrant yellow, cream-colored, or orange flowers; *E. capitatum* (western wallflower) is most widely distributed.

*Arabis* (rockcresses; see fig. d) are native perennials from rocky habitats, mostly with matted rosettes of leaves and racemes of purple, pink, or white flowers. The several species are identified by the position and shape of the siliques.

*Phoenicaulis cheiranthoides* (spear pod) is a native mountain perennial from rocky places, with basal rosettes of blue-green leaves and racemes of purple flowers followed by spear-shaped siliques.

TYPICAL GENERA WITH BROAD, ROUNDED SILICLES

NATIVES *Draba* (whitlow grass and others) are cushion-forming high-mountain perennials with yellow or white flowers and flattened, round silicles.

*Thysanocarpus* (lacepod; fringepod; see fig. e) are foothill annuals with minute white flowers and single-seeded, wheel-like silicles surrounded by a lacy or fluted rim.

*Lesquerella* (bladderpod) are perennials in rocky mountains, with flat rosettes of leaves, sprawling circles of stems, bright yellow flowers, and inflated, globe-shaped silicles.

ALIEN WEEDS *Capsella bursa-pastoris* (shepherd's purse), an annual with minute white flowers and triangular silicles shaped like an oldtime shepherd's purse.

*Lunaria annua* (honesty; money plant), a robust annual with purple flowers and flattened, coin-shaped silicles.

*Lobularia maritima* (sweet alyssum), a low perennial with fragrant white flowers and globe-shaped silicles; naturalized near the coast.

*Lepidium* (pepper grass; see fig. f), native and alien annuals and perennials with white or yellow flowers and round, flattened silicles with a minute notch at the top.

# BRASSICACEAE (Mustard Family)

### a. *Streptanthus tortuosus* (shield-leaf or common jewelflower)

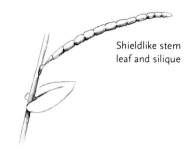

Shieldlike stem leaf and silique

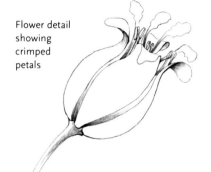

Flower detail showing crimped petals

### b. *Brassica rapa* (field mustard)

Cut-away flower closeup showing 4 sepals, 4 petals, and 6 stamens

### c. *Erysimum capitatum* (foothill wallflower)

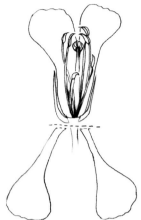

Cut-away view of flower showing 4 petals, 4 sepals, and 6 stamens

### d. *Arabis blepharophylla* (coast rockcress)

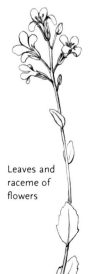

Leaves and raceme of flowers

Silique

Silique showing seeds inside

Flower detail

### e. *Thysanocarpus curvipes* (fringepod)

Single silicle surrounded by a wing

### f. *Lepidium* sp. (pepper grass)

Silicles showing intact fruits (top) and old fruits (bottom) with the false wall left behind

# CACTACEAE (Cactus Family)

**RECOGNITION AT A GLANCE** Plants with enlarged, fleshy green stems studded with clusters of spines at the nodes and producing (usually) showy, multi-petaled flowers with numerous stamens and an inferior ovary.

**VEGETATIVE FEATURES** Clump- and mound-forming stem succulents to woody succulent shrubs and small trees. Stems are often ribbed longitudinally or covered with prominent nipplelike "tubercles" at the nodes (areoles). Each areole features several to many radiating black, brown, yellow, white, or red spines.

**FLOWERS** Usually large and showy and are often arranged in a circle near the top crown of the stems. Flowers may be white, cream-colored, yellow, pink, rose-red, greenish, brownish, or other colors.

**FLOWER PARTS** Numerous spirally arranged sepals that grade into numerous petals, numerous separate stamens, and a single pistil with an inferior ovary and many, fingerlike stigma lobes.

**FRUITS** Dry to fleshy berries, often bright red or yellow when ripe. Fruits are covered with clusters of spines.

**RELATED OR SIMILAR-LOOKING FAMILIES** Few other families are likely to be confused with the cacti, which are easily identified by the combination of showy, multi-petaled flowers and highly modified, succulent stems with spines. Some confusion may exist with ocotillo (*Fouquieria splendens*) in the candlewood family (Fouquieriaceae), which has cactuslike stems lined with spines. However, the stems are not truly fleshy, the spines develop from single leaves, and the flowers are tubular, 5-petaled, and have a superior ovary. Ocotillo occurs in the Sonoran Desert and elsewhere.

**STATISTICS** Around 2,000 species confined—with one exception—to the New World. The range includes southern Canada, the United States, Mexico, Central America, and South America. The greatest diversity is in Mexico and the highlands of South America. Several species are epiphytes in tropical forests, but the majority favor deserts. Besides the many curious species that cactus-fanciers covet, the epiphytic orchid and Christmas cacti (*Epiphyllum* and *Zygocactus* spp.) are grown as houseplants and in protected patios. A few prickly-pears have sweet fruits made into jelly and pads that are cooked as vegetables. Peyote (and a handful of other hallucinogenic species) are ingested during Indian religious rituals.

## CALIFORNIA GENERA AND SPECIES

Cacti are highly modified stem succulents that live on rocky aprons in California's deserts or on dry hills in the southern mountains. The region has 9 genera with 29 native species and one introduced species, fewer than other parts of the Southwest.

### *GENERA WITH LONG SPINES, TINY GLOCHIDS, AND JOINTED STEMS*

*Opuntia* formerly included both chollas and prickly-pears, but is now confined to the latter, whose species have flattened, often vertically oriented stems. Chollas, with cylindrical stems, have now been moved to the genus *Cylindropuntia*. Most species of both genera occur in warm to hot deserts but a few are also common in coastal southern California. *Opuntia ficus-indica* (Indian fig), from Mexico, sometimes escapes from cultivation. It is widely grown for its edible fruits and pads (called *nopalitos*). Common species of both genera are listed below.

**PRICKLY-PEAR SPECIES** *Opuntia basilaris* (beavertail cactus) has bluish green pads, tiny spines, and pink flowers.

*O. chlorotica* (Mojave prickly-pear; see fig. a) grows into a shrublike plant with bright green pads, long and short spines, and yellow flowers.

*O. erinacea* (old man prickly-pear; see fig. b) has green pads, long and tiny spines, and pale yellow to pinkish orange flowers. It lives in the high desert.

*O. littoralis* (coast prickly-pear) has both straight and curved spines and yellow or yellow-orange flowers. Its fruits are noted for being succulent and fleshy. It occurs on coastal hills in southern California.

CHOLLAS *Cylindropuntia acanthocarpa* (buckhorn cholla) has multibranched stems and yellow-green flowers. It is widespread.

*C. bigelovii* (teddy bear or jumping cholla) has thick branches and spines (and the branches easily come loose) and reddish green flowers. It reproduces mostly by joints breaking off and taking root and is most typical of the Sonoran Desert.

*C. prolifera* (coast cholla) has a treelike appearance and features purple-red flowers. It is common along bluffs and scrub on the south coast.

GENERA WITH TALL, COLUMNAR STEMS
*Bergerocactus emoryi* (velvet cactus) forms colonies of stems with attractive brownish spines and pale yellow flowers. It occurs on the southernmost San Diego coast.

*Carnegiea gigantea* (giant saguaro; see fig. c) is a dramatic small tree to 50 feet high with upturned arms and evening-blooming, fragrant white flowers. Typical of Arizona's Sonoran Desert, it just manages to cross the Colorado River into eastern California.

GENERA WITH NIPPLELIKE TUBERCLES
*Mammillaria* (fishhook cacti) are small, moundlike cacti with at least one spine per node clearly hooked. Three native species with pink or white flowers.

*Escobaria* (beehive cacti) are similar in overall appearance to fishhhook cacti except all of the spines are straight, not hooked.

GENERA WITH RIBBED, BARREL-SHAPED TO CYLINDRICAL STEMS
*Ferocactus* (barrel cacti) stems are barrel shaped and lack woolly hairs; spines are flattened and pinkish. Two native species; *F. cylindraceus* (see fig. d) is common on rock aprons in our deserts; *F. viridescens* is found on far southern coastal bluffs and extends into Baja California.

*Echinocactus* (barrel cacti) are similar to *Ferocactus* except the stems are massed in dense clusters and are topped with woolly hairs. Our single species is *E. polycephalus* (cottontop).

*Echinocereus* (hedgehog cacti) have squat, rounded or cylindrical stems in large clusters and spiny fruits (fruits are spineless in the barrel cacti).

*E. engelmannii* (hedgehog cactus) has cylinder-shaped stems and magenta flowers.

*E. triglochidiatus* (mound cactus) has rounded stems and bright red flowers.

**a. *Opuntia chlorotica*
(Mojave prickly-pear)**

Plant habit

**b. *Opuntia erinacea*
(old man prickly-pear)**

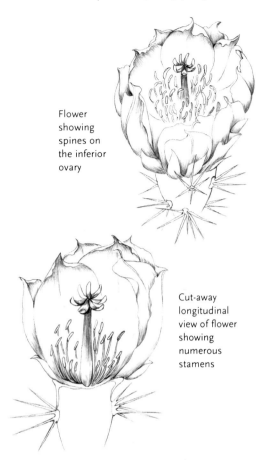

Flower showing spines on the inferior ovary

Cut-away longitudinal view of flower showing numerous stamens

**c. *Carnegiea gigantea* (saguaro)**

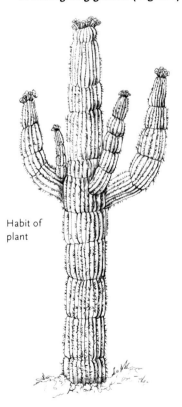

Habit of plant

**d. *Ferocactus cylindraceus* (barrel cactus)**

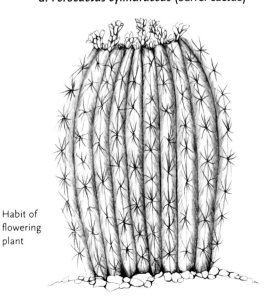

Habit of flowering plant

CACTACEAE (CACTUS FAMILY) 53

# CAMPANULACEAE (Bellflower Family)

**RECOGNITION AT A GLANCE** Herbaceous plants with alternate leaves, milky sap, and either bell-shaped, cup-shaped, or two-lipped flowers with an inferior ovary.

**VEGETATIVE FEATURES** Herbaceous annuals and perennials with simple, alternate leaves, and milky sap.

**FLOWERS** Bell-shaped, cup-shaped, or two-lipped; usually in racemes or single.

**FLOWER PARTS** 5 separate sepals, 5 irregular, 2-lipped or symmetrical petals joined to form a tube, 5 stamens attached to the tube, and a single pistil with an inferior, 1- to 3-chambered ovary and 2 or 3 stigmas.

**FRUITS** Capsules with numerous, tiny seeds. Capsules split along the side or open at the top.

**RELATED OR SIMILAR-LOOKING FAMILIES** Several other families feature two-lipped or bell-shaped flowers but few of those have an inferior ovary. The ovary together with the milky sap distinguish this family from others. Formerly, the family was split into Campanulaceae, with regular flowers, and Lobeliaceae, with irregular, two-lipped flowers.

**STATISTICS** Around 2,000 species widely distributed throughout the world. Many of the southern hemisphere species belong to different genera from those in the north. Several spectacular, treelike lobelioids occur in the mountains of central Africa and on the Hawaiian Islands. A number of ornamentals are grown in gardens including several bluebells (*Campanula* spp.) and lobelias (*Lobelia* spp.).

## CALIFORNIA GENERA AND SPECIES

The region has 10 native genera and several species.

*GENERA WITH REGULAR, AXILLARY FLOWERS*
*Heterocodon rariflorum* has small, cylinder-shaped flowers, some of which remain closed and are self-pollinated.

*Triodanis* (Venus's looking glass) has larger, saucer-shaped flowers. Both are uncommon annuals that appear after burns.

*GENERA WITH REGULAR FLOWERS CARRIED ABOVE THE LEAVES*
*Githopsis* (bluecups; see fig. a) are ephemeral annuals with seed pods shaped like an inverted cone and wrapped inside the sepals.

*Campanula* (bellflowers or bluebells; see fig. b) are annuals and perennials with oblong fruits that separate from the sepals.

*C. prenanthoides* (California harebell) is a common perennial in coniferous woods and blooms in summer.

*C. rotundifolia* (Scottish bluebell) is the classic bluebell, an uncommon perennial in northwestern forests.

*GENERA WITH IRREGULAR, TWO-LIPPED FLOWERS*
*Nemacladus* (threadstems) are ephemeral annuals from dry areas with threadlike stems and tiny, jewel-like flowers with unusual color combinations.

*Downingia* (downingias; see fig. c) are annuals living in temporary wetlands, with showy blue, purple, or white lobelialike flowers and a narrow, linear ovary. The various species are sorted out by the yellow and white color patches and purple spots on the lower lip.

*Lobelia* (lobelias) are perennials adapted to seeps and springs. Their flowers have globe-shaped ovaries.

*L. dunnii* (Dunn's lobelia) is a sprawling plant with small blue flowers, and lives in the mountains of south central and southern California.

*L. cardinalis* (cardinal flower) is a tall upright plant with showy, cardinal-red flowers, growing rarely in the mountains of southern California.

**a. *Githopsis pulchella*
(bluecup)**

**b. *Campanula scouleri*
(Scouler's harebell)**

Cut-away longitudinal view of flower (2 petals, sepals, and stamens removed for clarity)

Leafy stem with flowers

**c. *Downingia concolor* (Downingia)**

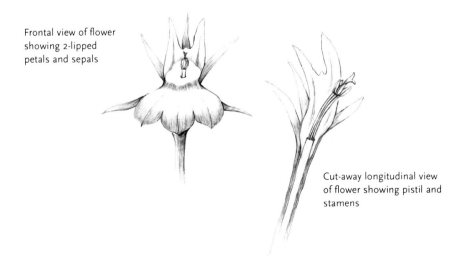

Frontal view of flower showing 2-lipped petals and sepals

Cut-away longitudinal view of flower showing pistil and stamens

CAMPANULACEAE (BELLFLOWER FAMILY)

# CAPPARACEAE (Caper Family)

**RECOGNITION AT A GLANCE** Plants with deeply divided to compound, ill-scented leaves and slightly irregular flowers with 4 petals and 6 stamens.

**VEGETATIVE FEATURES** Annuals, perennials, and small shrubs with deeply palmately lobed to compound, ill-scented leaves. Leaves may be toothed or not.

**FLOWERS** Small, slightly irregular, in spikelike clusters, and sometimes featuring long, spidery stamens.

**FLOWER PARTS** 4 sometimes partly joined sepals, 4 clawed petals, 6 stamens, and a single pistil with an ovary borne on a long stalk (stipitate).

**FRUITS** Capsules or sometimes nutlets.

**RELATED OR SIMILAR-LOOKING FAMILIES** The caper family is closely related to the mustard family (Brassicaceae) and may sometimes be confused with it. As of this writing, it is unclear just how these two families will eventually fall out, although it is likely the capers will be included as part of the larger mustard family. Major differences are that capers generally have distinctive and unpleasantly scented leaves, the flowers are often irregular, and the ovary is borne on a long stalk above the flower's receptacle. One mustard genus, *Stanleya*, (prince's plume) also has plants with stalked ovaries but their flowers are not irregular.

**STATISTICS** 800 species, mostly from arid and tropical areas. Besides the edible pickled buds of the Mediterranean caper plant (*Capparis spinosa*), the family is best known for the ornamental spider flowers (*Cleome* spp.).

## CALIFORNIA GENERA AND SPECIES

The region has six native genera and 12 species.

*Isomeris arborea* (bladderpod; see fig.) is a small shrub with bright yellow flowers and inflated, balloonlike seed pods.

*Oxystylis lutea* has axillary, headlike clusters of yellow flowers with spiny tails on their fruits.

*Polanisia dodecandra* (clammy-weed) has sticky hairs and white flowers with many stamens and a stalkless ovary (exceptions to the usual family characteristics).

*Wislizenia refracta* (jackass-clover) has yellow flowers and fruits lobed into 2 one-seeded nutlets.

*Cleomella* (little spider flower) has yellow flowers and capsules as broad as they are long.

*Cleome* (spider-flower) has yellow or rose-purple flowers and capsules shaped like tapered fuses—long and narrow. This and closely related genera are now considered to form a separate family, Cleomaceae.

### *Isomeris arborea* (bladderpod)

Flower detail

Whole fruit

Fruit cut open to show seeds

# CAPRIFOLIACEAE (Honeysuckle Family)

RECOGNITION AT A GLANCE Shrubs or small trees (mostly) with opposite leaves and no stipules, often bell-shaped to tubular flowers with inferior ovaries, and fleshy fruits.

VEGETATIVE FEATURES Deciduous shrubs, vines, or small trees with opposite, simple leaves, and (usually) no stipules.

FLOWERS Often bell-shaped or tubular; sometimes irregular with 2 lips, and arranged in complex cymes, spikes, or racemes.

FLOWER PARTS 5 minute sepals (hand lens), 5 petals fused to form a tube, 5 stamens attached to the tube, and a single pistil with an inferior, 2-chambered ovary.

FRUITS usually a fleshy berry, occasionally a capsule.

RELATED OR SIMILAR-LOOKING FAMILIES Although there are many shrubs with opposite leaves, few feature flowers with inferior ovaries and fleshy berries. Recent research has removed *Sambucus* (elderberry) and *Viburnum* and placed them in the family Adoxaceae. This is a logical change because both genera have complex cymes of tiny, shallow flowers unlike those of the rest of the Caprifoliaceae.

The honeysuckle family shares features in common with the mostly tropical coffee family (Rubiaceae), from which it is separated with difficulty. Differences include the presence of stipules and usually regular flowers in the Rubiaceae.

STATISTICS 450 species found primarily in the temperate part of the northern hemisphere and especially diverse in the mountains of China. There are a number of garden ornamentals including *Abelia*, *Weigela*, and various honeysuckles (*Lonicera* spp.). The elderberries and viburnums have been removed to the family Adoxaceae.

## CALIFORNIA GENERA AND SPECIES

The region has three native genera and 14 mostly native species. Elderberries and viburnum have brief mention at the end because of their long association with this family.

*Linnaea borealis* (twinflower) is a creeping, woody ground cover with pairs of nodding, pink, bell-shaped flowers. It lives in moist coniferous forests in the northwest.

*Symphoricarpos* (snowberries) have irregularly toothed and lobed leaves; hanging, bell-shaped pinkish flowers; and snowy white berries.

*S. albus* var. *laevigatus* (see fig. a) is an upright shrub in foothill woodlands.

*S. mollis* is a creeping ground cover from many wooded habitats.

*Lonicera* (honeysuckles) are woody vines and shrubs with tubular, usually 2-lipped flowers.

*L. hispidula* var. *vacillans*, *L. interrupta*, and *L. ciliosa* are vines with pale pink, pale yellow, and bright orange flowers respectively.

*L. subspicata* (southern honeysuckle) is a half-twining shrub with white flowers from southern California.

*L. involucrata* (twinberry honeysuckle; see fig. b) is a shrub with tubular yellow-orange flowers set in pairs of red bracts.

### ELDERBERRIES AND VIBURNUM

Currently the following genera are placed in the Adoxaceae because of their tiny, symmetrical flowers borne in complex panicles and cymes:

*Sambucus* (elderberries) are large shrubs to small trees with pinnately compound leaves and dense cymes of tiny whitish flowers.

*S. mexicana* (blue elderberry; see fig. c) has flat-topped clusters of flowers and is widespread in canyons inland and in the mountains.

*S. racemosa* (red elderberry) has pyramid-shaped clusters of flowers and is confined to moist coastal canyons.

*Viburnum ellipticum* (native viburnum) is an uncommon shrub with simple, toothed leaves and dense clusters of small white flowers from the north Coast Ranges and Sierra foothills.

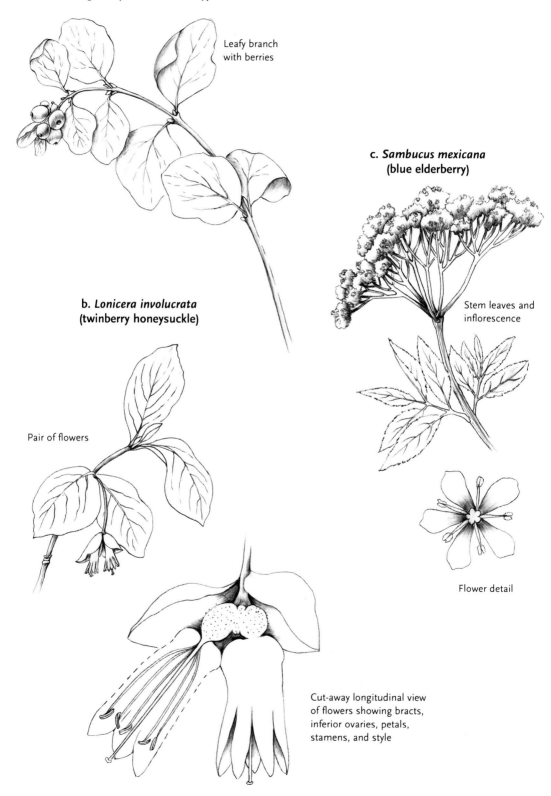

# CARYOPHYLLACEAE (Pink Family)

RECOGNITION AT A GLANCE Herbaceous plants with pairs of leaves on swollen nodes, and flowers with separate, clawed petals that are often notched or fringed.

VEGETATIVE FEATURES Herbaceous annuals and perennials with pairs of simple leaves attached to swollen nodes (run your finger down the stem). Some have papery stipules. Flowers vary from inconspicuous and petalless to showy with an open pinwheel–like design, and are arranged in various ways.

FLOWER PARTS 5 separate to partly fused sepals, 5 separate clawed petals often notched or fringed (pinked), 5 or 10 stamens, and a single pistil with a superior ovary and 3 to 5 styles and stigmas.

FRUITS Capsules with many seeds borne on a central stalk (free-central placentation) inside the ovary.

RELATED OR SIMILAR-LOOKING FAMILIES The pink family belongs to a special alliance of families with unique pigments (betalains) and other properties different from other dicots. This alliance includes the portulaca family (Portulacaceae), cactus family (Cactaceae), buckwheat family (Polygonaceae), and goosefoot family (Chenopodiaceae), most of which have floral traits that immediately distinguish them from the pinks. (Some of these families are not detailed in this book.)

STATISTICS Around 2,400 species widely distributed but not diverse in the tropics. The greatest diversity is in the mountains of the northern hemisphere and in the Mediterranean region. Several are grown as cherished garden flowers including carnation (*Dianthus caryophylla*), pinks (*Dianthus* spp.), crown-pink (*Lychnis coronaria*), snow-in-summer (*Cerastium tomentosum*), Irish-moss (*Sagina*), and various catchflies (*Silene* spp.).

## CALIFORNIA GENERA AND SPECIES

The region has 28 genera, several nonnative or with nonnative species.

## TYPICAL NATIVES

*Silene* (native pinks), several perennial species with fused, often sticky, cylinder-shaped sepals and fringed and appendaged petals.

*S. californica* (see fig. a) and *S. laciniata* (Indian pinks) have scarlet flowers and live in woods throughout the state.

*S. hookeri* (Hooker's pink) has floppy stems and very large pale pink flowers. It occurs in woods in the northwest.

*Cerastium arvense* (bluff or meadow chickweed) is a floppy-stemmed perennial with notched, snowy white petals. It lives on coastal bluffs and in mountain meadows. (A few cerastiums are introduced weeds with tiny or nonexistent petals.)

*Spergularia* (sand-spurreys) have fleshy leaves and small pale purple flowers. Some are native and others introduced. They live in hard-packed soils and rocky places.

*Minuartia* (sandworts) are mostly tiny annuals with linear leaves and white flowers.

*Arenaria* (also sandworts; see fig. b) are mostly clump-forming mountain perennials with narrow leaves and varied arrangements of white flowers.

*Sagina* (pearlworts) are tufted plants with narrow, often linear leaves and tiny white flowers.

*Stellaria* (chickweeds) are native perennials whose deeply lobed white petals resemble a 10-pointed star.

## COMMON NONNATIVES

*Stellaria media* (chickweed) is a sprawling annual with tiny white flowers, each petal deeply lobed to look like 2 petals.

*Silene gallica* (windmill pink) is an upright, hairy annual with ribbed sepals and pinwheel-like white to pale pink petals.

*Saponaria officinalis* (soapwort, bouncing bet) is a clump-forming perennial with broad leaves and dense clusters of showy, pale pink flowers.

*Spergula arvensis* (spurrey) is a sprawling annual with whorled linear leaves and tiny white flowers.

*Petroraghia prolifera* (tunic flower) is an upright annual with turban-shaped bracts around flowers with notched pink petals.

**a. *Silene californica* (Indian pink)**

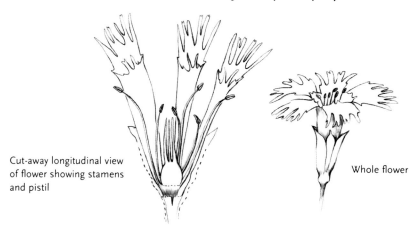

Cut-away longitudinal view of flower showing stamens and pistil

Whole flower

**b. *Arenaria kingii* (King's sandwort)**

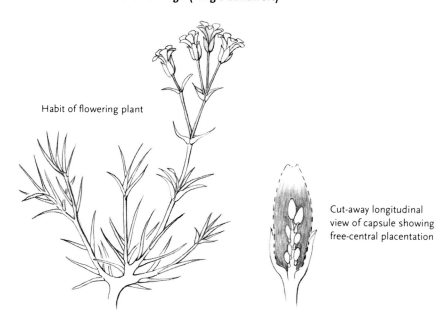

Habit of flowering plant

Cut-away longitudinal view of capsule showing free-central placentation

# CHENOPODIACEAE (Goosefoot Family)

Chenopodiaceae is now considered part of Amaranthaceae, the amaranth family.

RECOGNITION AT A GLANCE Plants with simple leaves often covered by matted hairs or scurfy scales. Flowers are tiny, inconspicuous, and petalless; fruits are sometimes surrounded by wings.

VEGETATIVE FEATURES Herbaceous annuals, perennials, and small shrubs with simple, linear to elliptical or triangular, gray to pale green leaves covered with dense, multibranched hairs or scurfy scales (use a good hand lens). In some genera the leaves are reduced to scales; several species have fleshy stems or leaves.

FLOWERS Minute, greenish, wind-pollinated, unisexual, and arranged in dense panicles.

FLOWER PARTS 5 separate sepals, no petals, 5 stamens, and a single pistil with a superior, 1-chambered ovary.

FRUITS Tiny, 1-seeded utricles (inflated achenes) often surrounded by 1 or more bracts that sometimes form wings.

RELATED OR SIMILAR-LOOKING FAMILIES The chenopods often have a striking presence although their flowers seldom call attention to themselves. The distinctive fruits are a good overall identifier when present. In vegetative features, some of the arid composites such as bursage (*Ambrosia* spp.) and spiny budscale (*Artemisia spinescens*) look superficially similar but are told apart by their lack of complex hairs and/or scurfy scales. Other members of the amaranth family (Amaranthaceae) present a similar appearance but lack the scurfy hairs of the chenopods, produce hard, 1-seeded fruits, and often have spiny bracts among the flowers. The two families were combined following investigation of DNA evidence.

STATISTICS 1,300 species and common indicators of salty soils. Many species dominate salt marshes, alkali flats, and shadscale scrub. Many species are important components of deserts around the world. Although they are key players in their communities, few are cultivated or used for food, although several species have edible leaves, seeds, or stems. The exceptions are quinoa (*Chenopodium quinoa*), a grain from the high Andes, and *Beta vulgaris*, which gives us chard, beets, and sugar beets.

## CALIFORNIA GENERA AND SPECIES

The region has 11 native genera; seven nonnative, introduced genera; and many species.

*PLANTS WITH FLESHY STEMS AND/OR LEAVES*
*Salicornia* (pickleweed) are herbaceous plants with opposite, scalelike leaves and pickle-shaped stems. They are major dominants in coastal salt marshes.

*Allenrolfea occidentalis* (iodine bush) looks similar to pickleweed but is a small shrub with alternate leaves. It lives in a variety of habitats in salty soils.

*Sarcobatus vermiculatus* (wormwood, greasewood) is a shrub with wormlike, fleshy leaves and winged fruits. It lives in salty desert flats.

*Suaeda* (sea-blite) is herbaceous or partly woody with fleshy leaves and wingless fruits. It lives in a variety of salty soils.

*PLANTS LACKING FLESHY STEMS OR LEAVES*
*HERBACEOUS PLANTS*
*Chenopodium* (lamb's quarters and others) has scurfy, nonspiny leaves. Many species are summer weeds in dry soils but a few species are native. *C. californicum* (soaproot) is a vigorous herbaceous perennial that lives in a variety of open habitats.

*Salsola tragus* (Russian-thistle or tumbleweed) is an aggressive introduced plant with smooth, spine-tipped leaves. The whole plant rounds up when dry and tumbles far across the land, spreading its fruits as it rolls.

SHRUBS *Atriplex* (saltbushes; see fig.) are herbaceous perennials and shrubs with scurfy leaves, and fruits enclosed in a pair of usually triangular bracts. Many species are widespread in dry salty habitats and deserts.

*Grayia spinosa* (hopsage) is a small shrub with softly hairy leaves and clasping, rounded bracts that are bright red to pink when ripe. It is widespread in dry areas of southern California.

*Krascheninnikovia lanata* (mulefat) is a small shrub with narrow, white leaves covered with dense, starlike hairs. It produces fruits covered with woolly hairs and is typically found in low, dry deserts.

**Atriplex hymeneletra (desert holly)**

Leaf and fruiting spike

Female flower detail

Fruit detail with bracts

Cluster of female flowers

# CONVOLVULACEAE (Morning Glory Family)

RECOGNITION AT A GLANCE Ground covers or vines with milky sap and (usually) large, funnel-shaped white, pale yellow, or pink flowers with completely fused petals.

VEGETATIVE FEATURES Free-living vines and ground covers with alternate, often round or arrowhead-shaped leaves, and milky sap.

FLOWERS Usually showy, funnel- or bell-shaped, white, pale yellow, or pink, and usually solitary. Sepal-like bracts sometimes occur just below the flower.

FLOWER PARTS 5 separate to partly joined sepals, 5 highly fused, pleated petals, 5 stamens attached to petals, and a single pistil with a superior, 2-chambered ovary.

FRUITS Capsules with a few large, hard seeds.

RELATED OR SIMILAR-LOOKING FAMILIES The flowers of the morning glory family often resemble the flowers in the nightshade family (Solanaceae), but possess several easily observable differences, including an often-vining habit, milky sap, funnel-shaped flowers, and few-seeded fruits. Nightshades are seldom viny; they have no milky sap; their flowers are usually flat, tubular, or bell-shaped; and their often fleshy fruits contain many seeds.

STATISTICS 1,000 species, especially diverse in the tropics but also in temperate parts of the northern hemisphere. Tropical species include shrub and tree forms that are missing from our flora. The South American sweet potato (*Ipomoea batatas*) is a major staple food in some parts of the world. Several ornamental vines, often aggressive in gardens, include morning glories (*Ipomoea* and *Convolvulus* spp.) and the lawn substitute *Dichondra*. Many have poisonous alkaloids.

## CALIFORNIA GENERA AND SPECIES

The region has five genera; 2 are mostly nonnative.

*Convolvulus arvensis* (bindweed) is a pernicious European weed with underground rhizomes, arrowhead-shaped leaves, and white to pinkish flowers.

*Calystegia* (wild morning glories; see fig.) are perennial ground covers or twining vines with round to arrowhead-shaped leaves, funnel-shaped pale yellow, white, or pink flowers. They are not invasive and live in many different habitats.

*Cressa truxillensis* (alkali weed) is a low-growing perennial with silvery leaves and small white flowers. It lives in salty soils near the coast.

*Dichondra repens* (dichondra) is a perennial, creeping ground cover with round leaves and tiny flowers. It lives in semishaded or grassy habitats in the foothills.

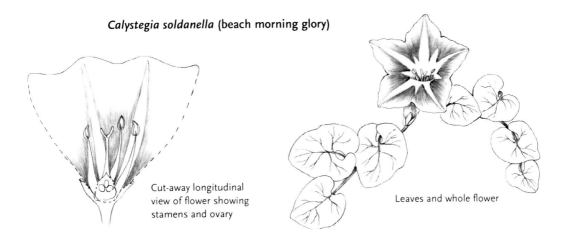

***Calystegia soldanella*** (beach morning glory)

Cut-away longitudinal view of flower showing stamens and ovary

Leaves and whole flower

# CORNACEAE (Dogwood Family)

**RECOGNITION AT A GLANCE** (Usually) deciduous shrubs or small trees with pairs of ovate leaves and tiny white or green, 4-petalled flowers with an inferior ovary.

**VEGETATIVE FEATURES** Deciduous shrubs, ground covers, or small trees with opposite, simple, ovate leaves and a pinnate-arcuate vein pattern. The paired new twigs are often red.

**FLOWERS** Small, star-shaped, white or green, and arranged in flat-topped cymes or buttonlike heads sometimes surrounded by white or pink, petal-like bracts.

**FLOWER PARTS** 4 minute sepals, 4 separate petals, 4 stamens, and a single pistil with an inferior ovary.

**FRUITS** Fleshy white, purple, or red drupes sometimes clustered together to form multiple fruits.

**RELATED OR SIMILAR-LOOKING FAMILIES** Although the dogwood family was sometimes considered related to other woody families such as the garrya family (Garryaceae) and honeysuckle family (Caprifoliaceae), the distinctively veined leaves and flower details easily separate them.

**STATISTICS** Around 100 species in the north temperate zone in wooded areas. Several dogwoods (*Cornus* spp.) are noted for their beautiful spring flowers and are widely cultivated. By some definitions, the family also includes other popular shrubs such as *Aucuba japonica* (gold dust plant) and *Corokia cotoneaster*.

## CALIFORNIA GENERA AND SPECIES

The region has one native genus with five species.

*Cornus nuttallii* (flowering or mountain dogwood; see fig. a) is a small tree with large, white, petal-like bracts around a head of greenish flowers. It lives in forests in the north Coast Ranges and Sierra Nevada.

*C. canadensis* (bunchberry) is a woody, creeping ground cover with similar bracts around a head of purplish flowers. It is uncommon in northwestern conifer forests.

*C. sericea* (red-twig dogwood; see fig. b), *C. glabrata* (brown dogwood), and *C. sessilis* (black-fruited dogwood) are vigorous, multi-trunked shrubs that live along wooded streams. The first two have white flowers in cymes; *C. sericea* has bright red new twigs; *C. glabrata* has brownish twigs. Black-fruited dogwood has umbel-like clusters of tiny green flowers, brown twigs, and purple-black fruits.

### a. *Cornus nuttallii* (flowering dogwood)

Leaves and flower cluster with petal-like bracts

Flower detail

### b. *Cornus sericea* (red-twig or creek dogwood)

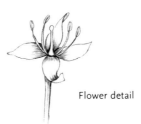

Flower detail

# CRASSULACEAE (Stonecrop or Live-Forever Family)

**RECOGNITION AT A GLANCE** Leaf succulents often with rosettes of leaves, cylinder-shaped or star-shaped flowers, and 5 partly joined to separate pistils.

**VEGETATIVE FEATURES** Succulent herbaceous perennials or annuals with simple entire, alternate, fleshy leaves arranged in rosettes or dense clusters.

**FLOWERS** Small, cylinder-shaped or star-shaped, and often in complex, flat-topped to candelabralike cymes.

**FLOWER PARTS** 5 usually separate fleshy sepals, 5 separate petals, 5 or 10 stamens, and 5 separate to partly joined pistils, each with a superior, 1-chambered ovary.

**FRUITS** Capsules (pistils partly joined) or follicles with numerous tiny, narrow, dustlike seeds.

**RELATED OR SIMILAR-LOOKING FAMILIES** Some other families have species with succulent leaves, the most prominent being the portulaca family (Portulacaceae), whose single pistil and different flower arrangements immediately separate it. The combination of the partly separate simple pistils and fleshy leaves usually identifies the stonecrop family.

**STATISTICS** 1,500 species found throughout the world but highly diverse in rocky habitats and deserts in North America (especially Mexico), the Canary Islands, Eurasia, and South Africa. These leaf succulents are often cultivated in rock gardens and containers for their handsome foliage, which is often blue-, gray-, or red-tinted, and for their attractive flowers. Common garden genera include *Echeveria*, *Sempervivum*, *Aeonium*, *Crassula*, *Sedum*, and *Kalanchoe*.

## CALIFORNIA GENERA AND SPECIES

The region has six genera; four mostly native and two nonnative.

### ANNUALS WITH TINY LEAVES

*Crassula* (pygmy weed) are plants seldom over an inch tall, with pairs of rounded leaves and minute, axillary, green or reddish flowers.

*Parvisedum* (annual stonecrops) are plants over an inch tall with alternate, yellow-green leaves and cymes of yellow, star-shaped flowers.

### PERENNIALS (USUALLY) WITH OBVIOUS LEAVES

*Sedum* (stonecrops) have leaves mostly in rosettes and star-shaped, yellow, white, or red flowers in cymes. The flowering stalk emerges from the middle of the rosettes. Sedums live in rocky habitats throughout the foothills and mountains: *S. spathulifolium* (common stonecrop; see fig. a) is most widespread and features gray, green, or red-tinted, spoon-shaped leaves in tight rosettes. *S. roseum* subsp. *integrifolium* (rosecrown; see fig. b), from the high mountains, has dark red flowers and goes dormant to a fleshy tuber.

*Dudleya* (dudleya, live-forevers) have leaves mostly in basal rosettes and yellow, red, orange, pink, or white flowers in candelabralike clusters. The flowering stalk emerges between the leaves of the rosette. There are many species especially in southern California.

*D. farinosa* (bluff lettuce) has chalky gray to bright green leaves and pale yellow flowers and lives along the north and central coasts.

*D. cymosa* (hot-rock dudleya; see fig. c) has gray-green leaves and pale yellow to red-orange flowers and lives on rock faces in the interior foothills.

a. *Sedum spathulifolium* (common stonecrop)

b. *Sedum roseum* subsp. *integrifolium* (rosecrown sedum)

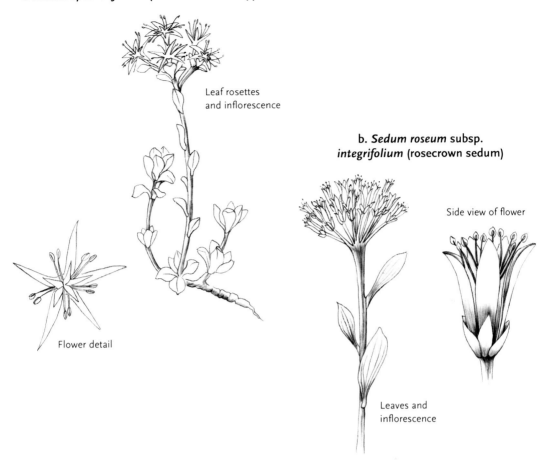

c. *Dudleya cymosa* (foothill or hot-rock dudleya)

# CUCURBITACEAE (Gourd or Cucumber Family)

**RECOGNITION AT A GLANCE** Vines with tendrils and unisexual flowers; female flowers with an inferior ovary that ripens into a exploding capsule or gourdlike pepo.

**VEGETATIVE FEATURES** Monoecious, perennial vines with curled tendrils, rounded, often palmately lobed leaves, and taproots or enormous tuberous roots.

**FLOWERS** Star-shaped or squashlike, yellow or white, unisexual, and arranged in racemes or single with the female flowers at the base of the raceme.

**FLOWER PARTS** 5 sepals, 5 partly fused petals, 5 fused stamens (male flowers), and a single pistil with an inferior ovary (female flowers).

**FRUITS** Berrylike pepos with a hard rind, or exploding capsules.

**RELATED OR SIMILAR-LOOKING FAMILIES** Few other families resemble the gourd family. The few others that are vines with tendrils include the pea family (Fabaceae), usually with pinnately compound leaves and very different-looking flowers; and the vine family (Vitaceae), with dense panicles of tiny, yellow-green flowers and grapelike fruits.

**STATISTICS** 700 species most diverse in the tropics, especially in tropical America. Besides the utility of gourds as containers, several important edible plants include watermelon from Africa, other melons from the Middle East, cucumbers from India, and squashes and pumpkins from Mexico and Central America. A few are also occasionally grown as ornamentals in gardens.

## CALIFORNIA GENERA AND SPECIES

The region has three native genera and three introduced genera.

*Marah* (Indian cucumbers; manroots) are rank vines from woodlands, forests, and chaparral with whitish flowers and explosive, usually prickly seed pods.

*M. fabaceus* (common manroot; see fig. a) has a wide distribution through the Coast Ranges and Sierra; its flowers are an off-cream color.

*M. oreganus* (Oregon manroot) lives in coastal forests and features larger leaves and pure white flowers.

*M. macrocarpus* (big-fruited manroot) lives in several plant communities in southern California and is noted for its very large, spiny fruits.

*Cucurbita* (calabazilla, coyote gourd) lives in arid foothills and deserts and has large, rough-textured leaves, yellow squashlike blossoms, and gourdlike fruits.

*C. foetidissima* (coyote gourd or melon; see figs. b and c) has unlobed leaves.

**a. *Marah fabaceus* (common manroot)**

**b. *Cucurbita foetidissima* (coyote gourd or calabazilla)**

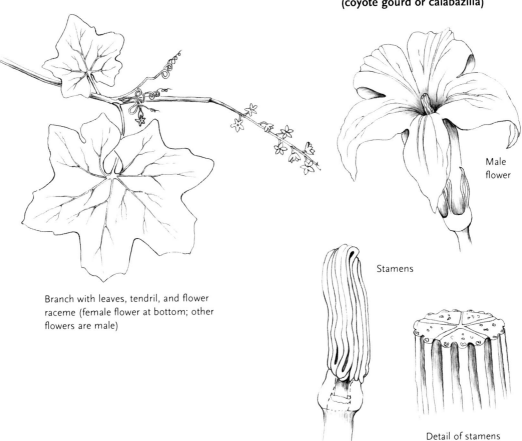

Male flower

Stamens

Detail of stamens

Branch with leaves, tendril, and flower raceme (female flower at bottom; other flowers are male)

**c. *Cucurbita foetidissima***

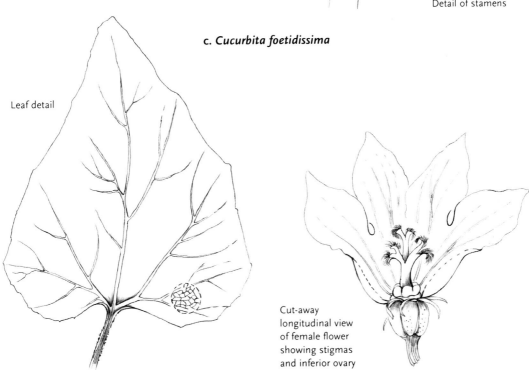

Leaf detail

Cut-away longitudinal view of female flower showing stigmas and inferior ovary

# CUPRESSACEAE (Cypress Family)

**RECOGNITION AT A GLANCE** Evergreen shrubs or trees with fragrant, scalelike leaves (most) and (usually) round to globe-shaped, woody or fleshy seed cones.

**VEGETATIVE FEATURES** Monoecious or dioecious, evergreen shrubs or trees with strips of fibrous bark and pairs or triplets of fragrant, scalelike leaves; juvenile leaves are spiny, awl-shaped needles.

Male (pollen) cones are tiny yellow nubbins freely produced on branches in summer and fall.

Female (seed) cones are usually egg- or globe-shaped with pairs or whorls of woody or fleshy scales.

**RELATED OR SIMILAR-LOOKING FAMILIES** Other conifer families are sometimes confused; the pine family (Pinaceae) features needlelike leaves and seed cones with spirally arranged scales. The redwood family (Taxodiaceae) usually has alternate, awl- to needle-shaped leaves and seed cones with spirally arranged scales. Some botanists propose combining the redwood and cypress families.

**STATISTICS** Around 120 species in temperate and arid climates of the northern and southern hemispheres but not in tropical lowlands. Some live in near-temperate rainforest conditions. The family is important for its fragrant, often rot-resistant wood, as, for example, western red-cedar (*Thuja plicata*) and Port Orford-cedar (*Chamaecyparis lawsoniana*). Many beautiful ornamental trees and dwarf cultivars are used in gardens from several different genera.

## CALIFORNIA GENERA AND SPECIES

The region has five native genera and 18 species.

### GENERA WITH FLATTENED, FERNLIKE BRANCHES

*Chamaecyparis lawsoniana* (Port Orford-cedar; Lawson cypress) has dark to bluish green foliage and round, marble-sized seed cones. It lives in the moist forests of the northwest.

*Thuja plicata* (western red-cedar) has dark to bright green foliage with bowtie-shaped, white stomatal bands underneath, and bell-shaped seed cones with pairs of scales. It is rare in moist forests in the northwest.

*Calocedrus decurrens* (incense-cedar; see fig. a) has bright yellow-green foliage without obvious stomatal bands, and seed cones that look like birds in flight, the upper scales curled up and out. It is a common component of mixed conifer forests in the mountains.

### GENERA WITH LEAVES IN THREE-DIMENSIONAL SPRAYS

*Cupressus* (recently moved to the New World genus *Callitropsis*, cypresses) vary from dwarf trees and large shrubs to trees with massive trunks and feature large, oval to globe-shaped woody seed cones that open after fire. California had a third of the world's species.

*C. macrocarpa* (Monterey cypress) is confined to windy bluffs on the Monterey peninsula but is widely planted elsewhere; it has horizontally trending, dark green branches.

*C. sargentii* (Sargent cypress; see fig. b) grows inland on serpentine soils through the Coast Ranges and features a rounded crown of pale to gray-green foliage.

*Juniperus* (junipers) vary from woody ground covers to trees with massive trunks; they feature fleshy, berrylike seed cones.

*J. communis* var. *saxatilis* (mat juniper; see fig. c) is a sprawling shrub that retains its sharp, juvenile, needlelike leaves for life. It lives in high-mountain rock scree.

*J. californica* (California juniper) is a small, multi-trunked tree of the hot inner foothills and desert mountains.

*J. occidentalis* and varieties (western or Sierra juniper) is a tree with a massive trunk and cinnamon-colored bark; it lives mostly in the high mountains.

### a. *Calocedrus decurrens* (incense-cedar)

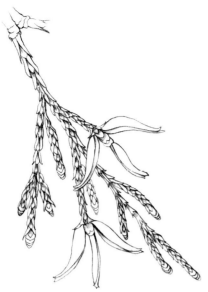

Scalelike leaves and 2 female seed cones

### b. *Cupressus sargentii* (Sargent cypress)

Scalelike leaves and seed cones

### c. *Juniperus communis* var. *saxatilis* (mat juniper)

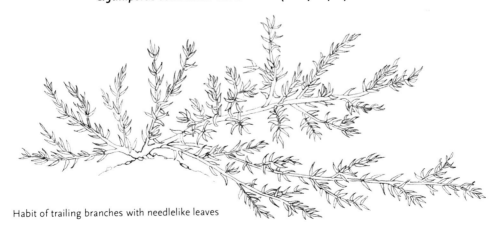

Habit of trailing branches with needlelike leaves

## CYPERACEAE (Sedge Family)

RECOGNITION AT A GLANCE Grasslike plants with channeled leaves arranged in 3 rows, (usually) triangular stems, and (often) separate spikes of male and female flowers.

VEGETATIVE FEATURES Perennial, usually rhizomatous, grasslike plants with solid, (usually) triangular stems and 3 rows of channeled leaves. Most have rough-textured leaves owing to silicon in the epidermis.

FLOWERS Uni- or bisexual, tiny, greenish or brownish, and usually arranged in spikelets, sometimes with the spikelets in umbels or heads.

FLOWER PARTS Perianth of 3 tiny scales or slender, often twisted bristles, 3 stamens, and a single pistil with a superior ovary.

FRUITS 1-seeded achenes sometimes enclosed in an inflated sac (perigynium).

RELATED OR SIMILAR-LOOKING FAMILIES Sedges are most often confused with the similar-looking grass family (Poaceae) and rush family (Juncaceae), although the three families are not considered to be closely related. Sedges differ from grasses by their solid, usually triangular stems, channeled leaves in 3 rows (not 2 as in grasses), and the presence of a highly modified perianth (grasses have none). Rushes differ from sedges by having usually flat or rounded, pithy stems, scalelike, irislike, or flattened leaves, and flowers with a starlike perianth of six tepals.

STATISTICS 3,600 species widespread throughout the world, often in moist to boggy habitats although some live in dry places. The wet growers often create the framework of mountain meadows and other high-elevation habitats, or border freshwater marshes. Several are serious weeds in California gardens. A few ornamental species, especially *Carex* species from New Zealand, add a bronze note to gardens. California Indians made extensive use of the strong fibers of native *Carex* rhizomes in weaving fine baskets.

### CALIFORNIA GENERA AND SPECIES

The region has 14 mostly native genera.

*Carex* (sedges; see fig. a) have spikelets of male flowers above the spikelets of female flowers, and the achenes are enclosed in a perigynium. The more than 130 species in the state live in a multitude of habitats. Most require microscopic examination of the fruits and flowers to identify the species.

*Cyperus* (nut-grass, umbrella sedge, and others; see fig. b) contain aggressive and weedy species as well as natives. They have bisexual flowers in complex arrangements surrounded by whorls of long, leafy bracts.

*Eriophorum gracile* (cotton-grass) is a mountain meadow plant with long, conspicuous white bristles that look like bolls of cotton.

*Eleocharis* (spike-rushes) are small to medium-sized plants with green stems replacing the scalelike leaves and terminal spikes of bisexual flowers.

*Scirpus* (bulrush, tule, and others) range from modest plants with grasslike leaves to very tall plants with green stems and scalelike leaves. The flower spikelets are arranged in heads or panicles.

*S. californicus* and *S. acutus* (tules) dominate freshwater marshes with stems to 10 feet high.

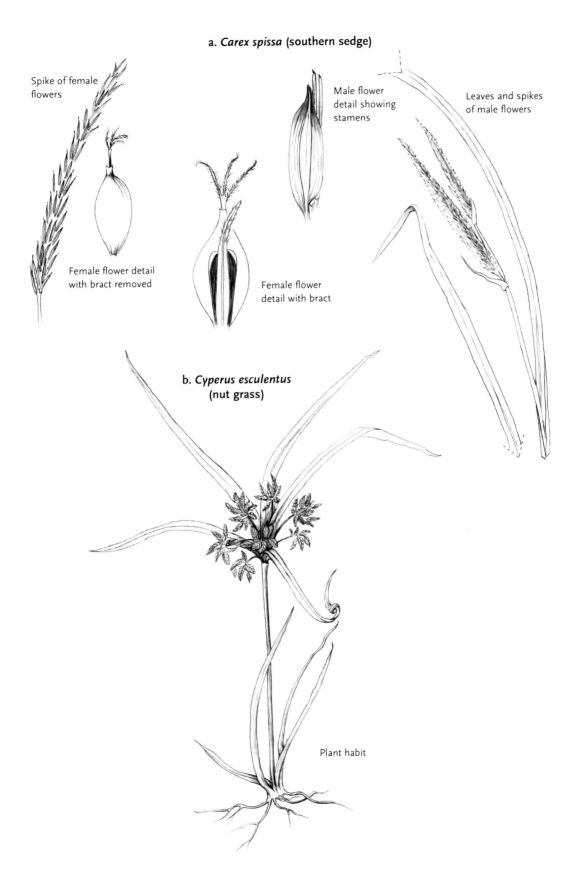

# EPHEDRACEAE (Joint-Fir Family)

RECOGNITION AT A GLANCE (Mostly) desert shrubs with jointed, green twigs, and leaves reduced to minute scales.

VEGETATIVE FEATURES Dioecious, small shrubs to 8 feet high with woody trunks, highly branched, green, jointed twigs (may also be bluish green), and pairs or triplets of tiny, scale-like leaves.

FLOWERS Lacking; these plants reproduce by tiny pollen and seed cones.

FLOWER PARTS Male and female cones are an inch or less long. Male or pollen cones have greenish bracts and protruding yellow, stamen-like pollen sacs. Female or seed cones have similar bracts and 1 to 3 seeds near their tips. Each seed features a tiny, soda-straw-like tip obvious at the time of pollination.

FRUITS Seed cones ripen 1 to few large, brown to black seeds.

RELATED OR SIMILAR-LOOKING FAMILIES The joint-fir family is unique in that it reproduces by cones rather than flowers. Yet these cones do not closely resemble the cones of conifers in the pine, cypress, and redwood families (Pinaceae, Cupressaceae, and Taxodiaceae). There are many other green-twigged desert shrubs, but few have the appearance of jointed twigs, and the minute leaves in Ephedraceae are always arranged in pairs or triplets.

STATISTICS A family with only 42 species distributed mostly in arid regions such as our Southwest, coastal South America, the Mediterranean region, and dry parts of China. Ephedra is noted for its possession of the medicinal compound ephedrine, used in small amounts to stimulate the heart but also misused in some diet pills. The twigs are steeped in hot water to make a tea in the American Southwest.

## CALIFORNIA GENERA AND SPECIES

One single genus, Ephedra (joint-fir, Mormon tea; see fig.), with seven native species. Most ephedras occur on sandy or rocky soils in the deserts. Species are separated on the basis of number of seeds per cone, color of twigs, and number of leaves per node.

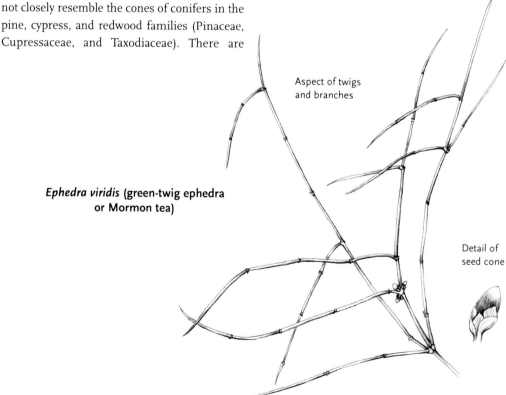

*Ephedra viridis* (green-twig ephedra or Mormon tea)

Aspect of twigs and branches

Detail of seed cone

# ERICACEAE (Heather Family)

## Ericoideae (Heather subfamily)

RECOGNITION AT A GLANCE Shrubs and trees with tough, simple leaves and bell-, urn-, tube-, or funnel-shaped flowers with joined petals and separate stamens. The anthers open by holes at their tips.

VEGETATIVE FEATURES Shrubs, woody ground covers, and trees with firm, simple, sometimes toothed, alternate leaves.

FLOWERS Showy, urn-, bell-, tube-, or funnel-shaped, and arranged in a variety of ways.

FLOWER PARTS (Usually) 5 sepals, (usually) 5 petals joined to form a tube or bell, 5 to 10 stamens attached to a nectar-bearing disc and not the petals, and a single pistil with a superior or inferior ovary. The stamens' anthers open by a pair of holes at their tips (not along their sides) and are often accompanied by hornlike appendages.

FRUITS Capsules with numerous, fine seeds, or berries with several seeds.

RELATED OR SIMILAR-LOOKING FAMILIES Most of the families with similar flowers such as the Pyrolaceae (wintergreen family) have been lumped with the heather family. Those species with bell- or urn-shaped flowers are unlikely to be confused with anything else.

STATISTICS Over 3,000 species found worldwide, with many genera diverse in specific geographical areas. Relatively few occur in tropical lowlands, but tropical cloud forests are home to many. The greatest number of species and genera prefer cool, moist, acid forests but some major genera have departed from that pattern and live in arid areas with Mediterranean climates. They include the manzanitas (*Arctostaphylos* spp.) in California and the heathers (*Erica* spp.) in South Africa. Other areas of diversity are the mountains of China and the Himalayas.

A great number of species are admired for their handsome foliage and beautiful flowers and are widely cultivated. Especially prominent are azaleas and rhododendrons (*Rhododendron* spp.), heathers, lily-of-the-valley shrub (*Pieris japonica*), and strawberry tree (*Arbutus unedo*). Fruits of huckleberries, blueberries, and cranberries (*Vaccinium* spp.) are widely grown for food.

## CALIFORNIA GENERA AND SPECIES

The region has 14 mostly native genera and one nonnative genus.

### GENERA WITH FLESHY FRUITS

*Gaultheria* (salal and others) are evergreen shrubs with zigzag stems, terminal, urn-shaped, white or pink flowers, and fleshy sepals that form a pseudo-berry around a papery ovary.

*G. shallon* (salal) is common in coastal conifer forests.

*Arbutus menziesii* (madrone) is a large, evergreen tree with orange-brown bark, magnolia-like leaves, trusses of urn-shaped white flowers, and warty, red-orange berries.

*Comarostaphylis diversifolia* (summer-holly) is a small, evergreen tree from far southern California with shreddy brown bark, broadly elliptical, toothed leaves; white, urn-shaped flowers; and warty, bright red berries.

*Arctostaphylos* (manzanitas; see fig. a) are woody ground covers, shrubs, and small trees with polished red bark; elliptical, evergreen leaves; trusses of urn-shaped, white to pink flowers; and reddish berries with stonelike seeds. The several dozen species live in many different habitats; California is the center of diversity for the genus.

*Vaccinium* (huckleberry, bilberry) are creeping to upright deciduous or evergreen shrubs with bell-shaped, white to pink flowers, and edible berries developed from an inferior ovary.

*V. ovatum* (evergreen huckleberry) is common in the understory of coastal conifer forests.

*V. parvifolium* (red huckleberry) is a deciduous species common in north coastal conifer forests.

### GENERA WITH DRY CAPSULE-TYPE FRUITS

*Rhododendron* (azalea, rosebay) are large shrubs with elliptical leaves, large, funnel-shaped flowers, and capsules filled with dustlike seeds.

*R. occidentale* (western azalea, fig. b) is a deciduous shrub with fragrant, white to pink flowers living along wooded watercourses.

*R. macrophyllum* (California rosebay) is an evergreen shrub with unscented, rose-purple flowers living on the edge of northern conifer forests.

*Ledum glandulosum* (Labrador tea) is an evergreen shrub from moist conifer forests with evergreen, azalealike leaves and clusters of small, white, saucer-shaped flowers.

*Kalmia polifolia* subsp. *microphylla* (bog laurel) is a sprawling shrub with tightly curled-under leaves and clusters of puckered, rose-pink flowers—each flower has 10 nectar pockets. It and the next two genera live in high mountain meadows.

*Cassiope mertensiana* (white-heather) is a mounding shrub with braided leaves and nodding, bell-shaped white flowers trimmed with pink sepals.

*Phyllodoce* (red-heathers) are low, colonizing shrubs with needlelike leaves and saucer- to bell-shaped, magenta flowers.

## Monotropoideae Subfamily

Monotropoideae is sometimes referred to as the ghost pipe subfamily.

RECOGNITION AT A GLANCE Herbaceous plants with green, variegated, or nongreen, scalelike leaves; racemes of urn-, saucer-, or openly bell-shaped flowers; and seed pods with thousands of minute, dustlike seeds.

VEGETATIVE FEATURES Herbaceous perennials, ground covers, fungus parasites with roots intimately connected to and dependent on mycorrhizal fungi.

FLOWERS Often waxy, urn-, bell-, or saucer-shaped, white, pink, or red, and single or arranged in racemes.

FLOWER PARTS The number of parts is similar to the Ericoideae (described above) but some species have stamens that open by lengthwise slits rather than pores at the ends.

FRUITS Capsules with thousands of tiny, dustlike seeds.

STATISTICS Perhaps a hundred or so species widely distributed on humus-laden conifer forest floors in the northern hemisphere. Most are impossible to cultivate because of their need for specific fungal partners, and so the subfamily has no economic value. These species have a remarkable relationship between fungi in the leaf litter layer on the forest floor and roots of conifers in the forest. The conifers produce sugar absorbed by the fungi, some of which are passed on to the roots of these parasitic and semiparasitic members of the Ericaceae.

## CALIFORNIA GENERA AND SPECIES

The region has 11 native genera.

### GENERA WITH GREEN LEAVES

*Pyrola* (wild wintergreen) has basal rosettes of broad, rounded leaves and racemes of flowers with a J-shaped style and stigma.

*P. picta* (white-veined shinleaf) has dark green leaves with conspicuous white veins and white flowers.

*P. asarifolia* (rose pyrola) has solid green leaves and pale pink-purple flowers.

*Orthilia secunda* (one-sided wintergreen) is similar to pyrola but has hanging, one-sided spikes of whitish green flowers.

*Monses uniflora* (woodland nymph) has small rosettes of wavy green leaves and 1 or 2 pure white, horizontally held, saucer-shaped flowers.

*Chimaphila* (pipsissewa, prince's pine; see fig. c) are semierect ground covers with dark green, lance-shaped leaves, and saucer-shaped, waxy pink or white flowers.

### GENERA WITH SCALELIKE LEAVES AND NO CHLOROPHYLL

GENERA WITH PARTLY JOINED PETALS AND URN-SHAPED FLOWERS *Pterospora andromeda* (pinedrops) has sticky, red-tinted stems and nodding white and brownish flowers.

*Sarcodes sanguinea* (snowplant; see fig. d) has stout, bright red stems and nodding, blood-red flowers. (The botanical name means *bloody flesh*.)

GENERA WITH SEPARATE PETALS *Allotropa virgata* (sugarsticks or candy cane) has pink-striped stems and outward-facing, saucer-shaped flowers with dark red stamens.

*Monotropa* (Indian or ghost pipe) has white stems and one or few nodding, pipe-shaped, white flowers.

### a. *Arctostaphylos manzanita* (common manzanita)

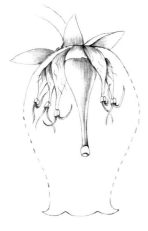

Cut-away longitudinal view of flower showing stamen details (some stamens have been removed for clarity)

Leaves and inflorescence

### b. *Rhododendron occidentale* (western azalea)

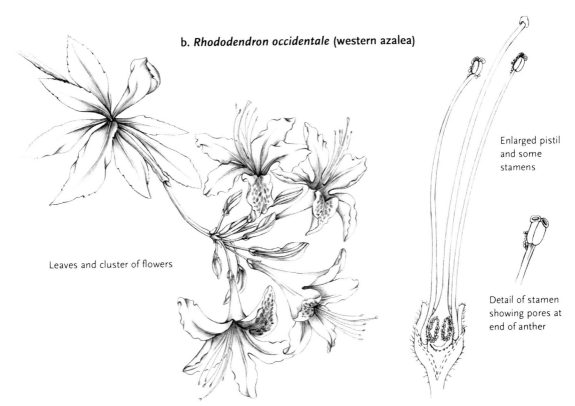

Leaves and cluster of flowers

Enlarged pistil and some stamens

Detail of stamen showing pores at end of anther

ERICACEAE (HEATHER FAMILY)

## ERICACEAE (Heather Family)

## MONOTROPOIDEAE (GHOST PIPE SUBFAMILY)

### c. *Chimaphila umbellata* (pipsissewa)

Face view of flower

### d. *Sarcodes sanguinea* (snowplant)

Habit of flowering plant

## EUPHORBIACEAE (Spurge Family)

RECOGNITION AT A GLANCE Plants often with a copious milky or caustic sap; tiny, unisexual, (usually) petalless flowers sometimes surrounded by colorful bracts and/or petal-like nectar glands; and a 3-lobed, superior ovary.

VEGETATIVE FEATURES Annual or perennial herbs and small shrubs sometimes with succulent stems. Simple linear, lance-shaped, spoon-shaped, rounded, or elliptical, sometimes toothed leaves.

FLOWERS Tiny, unisexual, and borne mostly in spikes, small clusters, or in flowerlike cyathia.

FLOWER PARTS 5 sepals (missing in *Euphorbia*), (usually) no petals, 1 to 5 or more stamens, and a single pistil with an (often) elevated, strongly 3-lobed ovary, and 3 prominent, often forked stigmas. In the large genera *Euphorbia* and *Chamaesyce*, the flowers are arranged in small cyathia that consist of a cup-shaped receptacle holding several male flowers of 1 stamen each and a single female flower with a stalked ovary. Cyathia are surrounded by petal-like nectar glands and often surrounded by varied, often colored bracts.

FRUITS Mostly explosive, 3-chambered capsules containing a few large seeds.

RELATED OR SIMILAR-LOOKING FAMILIES Although the spurge family is highly diverse, there are few other groups that resemble it. The combination of copious milky sap typical of many species, the unique cyathia of the spurges, and the prominent, 3-lobed ovaries serve to identify most euphorbs.

STATISTICS 7,500 species found throughout the world, with the greatest diversity in the tropics and arid areas. Habits range from tiny, weedy annuals to massive rain-forest trees; several African euphorbias resemble cacti. Garden ornamentals include perennial euphorbias, poinsettia (*Euphorbia pulcherrima*), the tropical crotons (*Codiaeum* spp.), chenille plants (*Acalypha* spp.), castor bean (*Ricinus communis*), and many more. The yuca or manioc (*Manihot esculenta*) is an important staple root crop in tropical America and Africa; the candlenut's (*Aleurites moluccana*) oily nuts fuel torches. Many species are noted for their poisons.

### CALIFORNIA GENERA AND SPECIES

The region has 10 partly to fully native genera with many species, and two introduced genera with two species.

*GENERA WITH CYATHIA-TYPE FLOWER ARRANGEMENTS*

*Euphorbia* (spurge; see fig. a) usually has alternate leaves and copious poisonous, milky sap. Many species are introduced and a few are native.

*Chamaesyce* (prostrate spurge) has opposite leaves and a similar sap. Many species are prostrate. Most species are native but a few are introduced garden weeds.

*GENERA WITHOUT CYATHIA*

SHRUBBY GENERA *Ricinus communis* (castor plant) is a highly poisonous plant with dramatic, palmately lobed leaves. It is commonly naturalized in coastal southern California.

*Croton californica* (California croton) is a sand dune plant with semisprawling stems; untoothed, densely white-hairy leaves; and yellowish flowers.

*Bernardia myriophylla* is an uncommon desert shrub with scalloped, hairy leaves and yellow-green flowers.

*Acalypha californica* (California copper-leaf) is a shrub from the western Sonoran Desert and mountains of southernmost California with ovate, inconspicuously hairy leaves and reddish flowers.

*Tetracoccus* are dioecious desert or southern-mountain shrubs with nearly hairless leaves of varied shapes and arrangements.

HERBACEOUS GENERA *Eremocarpus setigerus* (turkey mullein; see fig. b), now included in the genus *Croton*, is a mound-forming summer annual common along dry roadsides and has fragrant leaves covered with stiff, sometimes stinging, white hairs.

*Tragia stylaris* is a desert perennial that has leaves covered with stinging hairs and flowers in spikes with 3 to 6 stamens each.

*Stillingia* are desert perennials with hairless leaves and flowers in narrow spikes with 1 stamen each.

*Ditaxis* are desert perennials with hairy leaves and spikes of flowers with tiny, straw-colored petals.

**a. *Euphorbia peplus* (lesser spurge)**

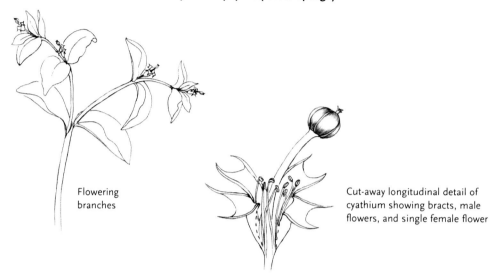

Flowering branches

Cut-away longitudinal detail of cyathium showing bracts, male flowers, and single female flower

**b. *Eremocarpus setigerus* (turkey mullein)**

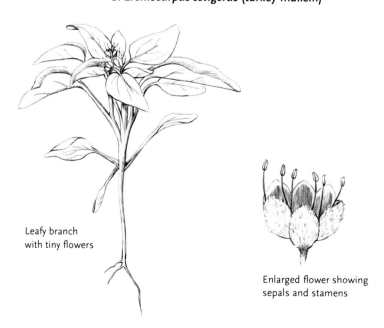

Leafy branch with tiny flowers

Enlarged flower showing sepals and stamens

# FABACEAE (Leguminosae; Pea or Legume Family)

**RECOGNITION AT A GLANCE** Usually compound leaves with stipules and flowers mostly similar to the sweetpea or acacia, followed by a beanlike seed pod (legume).

**VEGETATIVE FEATURES** Annuals, perennials, shrubs, and small trees with nitrogen-fixing nodules on the roots, and alternate, often compound leaves with stipules. Stipules are sometimes modified into glands or spines. Leaves may be trifoliate, palmately compound, or pinnately compound.

**FRUITS** Peapodlike, single-chambered legumes with (usually) 1 row of large seeds. Seeds are often laced with poisons as defense mechanisms against being eaten.

**RELATED OR SIMILAR-LOOKING FAMILIES** Despite the variation in flower design within the family, the pea family stands by itself. The legume-type fruit is unique to this family as is the flower design in each subfamily.

**STATISTICS** A huge, prominent family with 18,000 species worldwide and important in most floras including a vast number in Australia, where *Acacia* has several hundred species. The peas are second in economic importance to the all-important grass family (Poaceae); besides the ability of the nitrogen-fixing nodules to enrich soils, the family is a source of highly nutritious fodder for livestock (alfalfa, vetches, sweet-clover, and clovers), and of protein-rich seeds for the human diet (lentils, peas, beans, soy beans, and more). In addition, there are many notable temperate and tropical ornamentals too numerous to list. Prominent in California gardens are sweet pea, wisteria, hardenbergia, acacias, redbud, and many more.

The family is divided into three subfamilies, each with its own unique floral pattern, as follows.

## *Mimosoideae (Mimosa Subfamily)*

**FLOWERS** Tiny, yellow, pink, or white and massed into heads or spikes.

**FLOWER PARTS** 5 separate sepals; 5 tiny, separate petals; 5 to many long, colorful stamens; and a single pistil with a superior, 1-chambered ovary.

### CALIFORNIA GENERA AND SPECIES

The region has three native or partly native genera and one nonnative genus found primarily in deserts.

Acacias (*Acacia* spp.) are shrubs and trees with yellow or white flowers and often naturalized along the coast. Most of the ornamental species come from Australia. Our single native species, catclaw (*A. greggii*; see fig. a), occurs in the deserts of southern California.

## *Caesalpinoideae (Senna Subfamily)*

**FLOWERS** Showy, wide open, slightly irregular (usually), and often yellow.

**FLOWER PARTS** 5 separate sepals, 5 showy, irregular petals, 5 or 10 stamens, and a single pistil with a superior, 1-chambered ovary.

### CALIFORNIA GENERA AND SPECIES

The region has five mostly native genera and one nonnative genus; only one is native outside deserts.

*Cercis occidentalis* (western redbud; see fig. b) is a large, deciduous, foothill shrub with heart-shaped leaves and masses of red-purple, almost sweetpealike flowers.

## *Papilionoideae (Pea Subfamily)*

**FLOWERS** Showy, irregular and papilionaceous, in many colors, and arranged in spikes and racemes.

**FLOWER PARTS** 5 partly joined, sometimes irregular sepals, 5 petals with an upper back banner, 2 side wings, and 2 central joined petals that form a boat-shaped keel containing 5 or 10 sometimes joined stamens, and a single pistil with a superior, 1-chambered ovary. The form of the petals is called *papilionaceous*.

## CALIFORNIA GENERA AND SPECIES

The region has around 41 genera in all, several with some nonnative species or entirely nonnative species.

### NATIVE SHRUBS

*Pickeringia montana* (chaparral pea) is a small, thorny, chaparral shrub with trifoliate leaves and masses of magenta flowers.

### INTRODUCED TREES AND SHRUBS

*Genista, Cytisus,* and *Spartium junceum* (various brooms) are Mediterranean shrubs with green, broomlike twigs; small leaves; and racemes of bright yellow (occasionally white) flowers. They have become serious pests in native woodlands.

*Ulex europea* (gorse) is a broomlike shrub with leaves modified into sharp green spines and bright yellow flowers. It is especially widespread in coastal areas.

*Robinia pseudoacacia* (black locust) is an occasionally naturalized, deciduous tree with dark bark, pinnately compound leaves, thorns, and racemes of white flowers.

### VINY HERBACEOUS GENERA

*Lathyrus* (wild sweetpea) has pinnately compound leaves ending in tendrils and racemes of red, yellow, white, purple, or pink flowers. Under a hand lens, the style resembles a toothbrush. Several species from a variety of habitats; a few are introduced from Europe.

*Vicia* (vetch) has similar leaves and racemes of blue, purple, reddish, or pink flowers. The style ends in a stigma with a tufted ring of hairs. Several introduced species and two natives:

*V. americana* (common vetch) is widespread in wooded and grassy areas and features pink-purple flowers.

*V. gigantea* (giant vetch) is a vigorous vine in coastal woods and has dull red and greenish flowers.

### HERBACEOUS GENERA WITH PALMATELY COMPOUND OR TRIFOLIATE LEAVES

*Lupinus* (lupines; see fig. c) are annuals, perennials, and small shrubs with palmately compound leaves and spikelike racemes of blue, purple, pink, white, or yellow flowers.

*Pediomelum californicum* (California breadroot) has a tuberous root, palmately compound leaves, and headlike clusters of pale purple flowers.

*Thermopsis macrophylla* (false lupine) is a rhizomatous perennial with large, trifoliate leaves, conspicuous stipules, and racemes of bright yellow, lupinelike flowers.

*Melilotus* (sweet-clovers) are introduced annuals with trifoliate leaves and narrow spikes of white or yellow flowers.

*Medicago* (alfalfa; bur-clovers) are introduced annuals with trifoliate leaves and purple or yellow flowers and coiled, often spiny fruits that catch on clothing.

*Trifolium* (clovers) are nonnative and native annuals and perennials with (usually) trifoliate leaves and headlike clusters of small white, pink, red, purple, or yellow flowers.

### GENERA WITH PINNATELY COMPOUND LEAVES

*Lotus* (lotus) are highly variable annuals and perennials with off-center, pinnately compound or trifoliate leaves and axillary or umbel-like clusters of white, pink, red, or yellow flowers.

*L. scoparius* (deerbroom) is a green-twigged subshrub with bright yellow flowers that prospers after burns.

*L. crassifolius* (bull lotus) is a stout mountain perennial with substantial leaves and dark red flowers.

*Glycyrrhiza lepidota* (wild licorice) is a stout perennial with spikes of whitish flowers and spine-covered fruits.

*Astragalus* (loco-weeds, milk-vetches, rattle-pods; see fig. d) is a very large genus with over 80 species in California. It includes annuals and perennials with spikelike racemes of white, purple, pink, yellowish, or red flowers followed by characteristic inflated seed pods of various shapes. Many species are highly toxic; some accumulate selenium from soils. Identification of species requires ripe fruits.

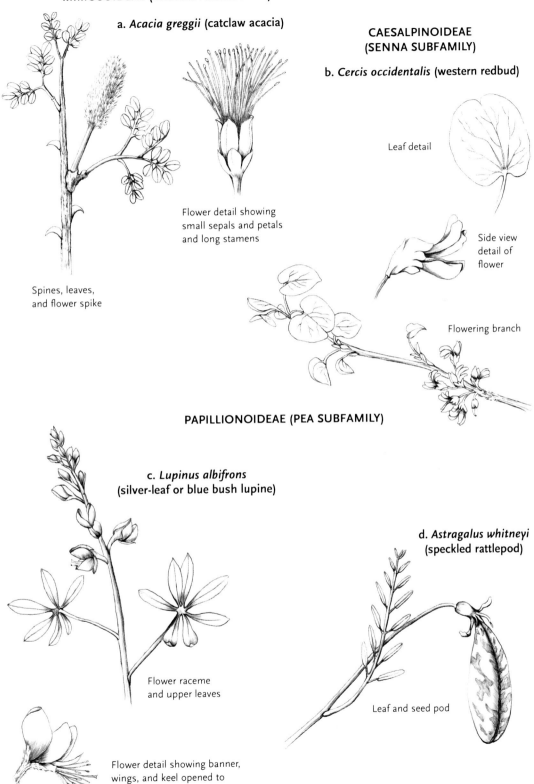

**MIMOSOIDEAE (MIMOSA SUBFAMILY)**

a. *Acacia greggii* (catclaw acacia)

Spines, leaves, and flower spike

Flower detail showing small sepals and petals and long stamens

**CAESALPINOIDEAE (SENNA SUBFAMILY)**

b. *Cercis occidentalis* (western redbud)

Leaf detail

Side view detail of flower

Flowering branch

**PAPILLIONOIDEAE (PEA SUBFAMILY)**

c. *Lupinus albifrons* (silver-leaf or blue bush lupine)

Flower raceme and upper leaves

Flower detail showing banner, wings, and keel opened to reveal stamens and style

d. *Astragalus whitneyi* (speckled rattlepod)

Leaf and seed pod

FABACEAE (LEGUMINOSAE; PEA OR LEGUME FAMILY)

# FAGACEAE (Oak or Beech Family)

**RECOGNITION AT A GLANCE** Shrubs or trees with simple or pinnately lobed leaves, petalless catkins of male flowers, and cups or spiny burs containing an acorn or nuts.

**VEGETATIVE FEATURES** Monoecious, evergreen or deciduous shrubs and trees with alternate, simple, and usually toothed or pinnately lobed leaves and deciduous stipules.

**FLOWERS** Unisexual, tiny, greenish or white; the males in stiff upright catkins or slender, hanging catkins. The female flowers are single or in small clusters and found at the base of the male catkins or in axils of the leaves on new shoots.

**FLOWER PARTS** The male flowers have bracts and a few stamens; the female flowers have spiny bracts or sit in a scaly cup with a few sepals and a single pistil and an inferior ovary with a 3-lobed stigma.

**FRUITS** Small nuts enclosed in spiny burs or acorns that sit in a cup of warty or scalelike bracts.

**RELATED OR SIMILAR-LOOKING FAMILIES** Other woody plant families with hanging male catkins include the birch family (Betulaceae) and garrya family (Garryaceae). The birches, however, have female catkins (usually) and tiny winged achenes except for hazelnut, which has a nut enclosed inside a fuzzy husklike set of bracts. The garryas have pairs of tough, untoothed leaves and female flowers in catkins on separate plants. (The garrya family is not detailed in this book.)

**STATISTICS** Upwards of 900 species mainly across the northern hemisphere but with the southern hemisphere genus *Nothofagus* (southern beech) from South America, New Zealand, Australia, the Pacific Islands, and New Guinea. The family is important for its many imposing trees that dominate woodlands and forests; many are considered keystone species in their habitats. Besides providing many fine shade trees and trees for timber, the chestnuts (*Castanea* spp.) and sometimes beechnuts (*Fagus* spp.) are used for food. The American Indians made wide use of leached oak acorns for food.

## CALIFORNIA GENERA AND SPECIES

The region has three native genera.

*Chrysolepis* (chinquapin) are shrubs or trees with untoothed leaves backed by tawny or golden scales, with white, candlelike male catkins, and nuts enclosed in spiny burs.

*C. sempervirens* (mountain chinquapin) is a mountain shrub.

*C. chrysophylla* (coast chinquapin; see fig. a) is a shrub or tree in coastal chaparral and forests.

*Lithocarpus densiflorus* (tanbark-oak) is a shrub or tree with large, toothed, evergreen leaves bearing a herringbone pattern of veins; white catkins of male flowers; and an acorn in a fringed cup.

*Quercus* (oaks) are shrubs or trees with variable deciduous or evergreen leaves; hanging, yellow male catkins; and acorns in a scaly or warty cup.

### WHITE OAKS

White oaks have pale bark and acorns that ripen in one year.

*Quercus garryana* (Garry oak) and *Q. lobata* (valley oak) have deeply pinnately lobed leaves. Valley oak is among our largest oaks and favors valley bottoms; Garry oak lives in hills in northern California and is widely scattered in the Sierra.

*Q. douglasii* (blue oak) has shallowly lobed, bluish leaves and lives in the hot, dry foothills around the Central Valley.

*Q. engelmannii* (Engelmann oak) has entire, bluish leaves, is semievergreen, and lives in the mountains of southern California.

Several white oaks are shrubby (scrub oaks); especially widespread in chaparral is *Q. berberidifolia* (common scrub oak).

### BLACK OAKS

Black oaks have dark bark and acorns that usually take two years to ripen.

*Quercus kelloggii* (California black oak) has deciduous, deeply pinnately lobed leaves with bristle-tipped lobes. It lives in uplands and middle elevations in the mountains.

*Q. agrifolia* (coast live oak; see fig. b) has cupped, evergreen leaves and lives in the Coast Ranges and at lower elevations of the mountains in southern California.

*Q. wislizenii* (interior live oak) has evergreen, flat leaves and lives in the hot interior foothills around the Central Valley.

## GOLDEN OAKS

Golden oaks have features intermediate between white and black oaks.

*Quercus chrysolepis* (canyon live or goldcup oak) has evergreen leaves that are pale whitish underneath and green above, and acorns in a warty cup with gold powder. It lives in many foothill and middle mountain habitats.

*Q. vaccinifolia* (huckleberry oak) is a dense, evergreen shrub with shiny, huckleberry-shaped leaves. It is found in rocky habitats in the high mountains.

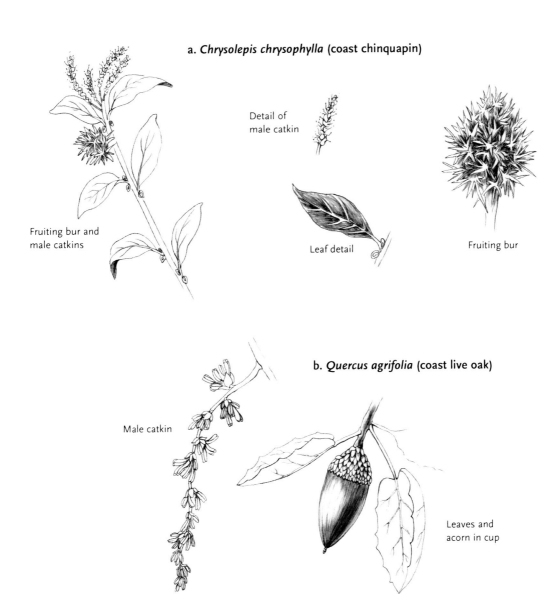

**a.** *Chrysolepis chrysophylla* (coast chinquapin)

Fruiting bur and male catkins

Detail of male catkin

Leaf detail

Fruiting bur

**b.** *Quercus agrifolia* (coast live oak)

Male catkin

Leaves and acorn in cup

FAGACEAE (OAK OR BEECH FAMILY)

# GENTIANACEAE (Gentian Family)

RECOGNITION AT A GLANCE Herbaceous plants with pairs of simple, usually elliptical, entire leaves and flowers with pleated petals that often have appendages and/or glands.

VEGETATIVE FEATURES Herbaceous annuals and perennials with pairs of simple, elliptical or broader, entire leaves, each leaf pair often at right angles to the next.

FLOWERS Usually showy, cup-shaped or starlike, and solitary or in various other, sometimes complex arrangements.

FLOWER PARTS 4 or 5 partly fused sepals, 4 or 5 pleated petals fused into a tube and often with appendages or conspicuous raised glands, 4 or 5 stamens joined to the tube, and a single pistil with a superior, partially 2-chambered ovary and usually 2-lobed stigma.

FRUITS Capsules with many seeds.

RELATED OR SIMILAR-LOOKING FAMILIES The key features of gentian flowers might suggest a few other families such as the waterleaf family (Hydrophyllaceae) and nightshade family (Solanaceae) but the combination of flower shape, appendages or prominent glands, and precisely aligned, opposite leaves easily separates it.

STATISTICS Around 900 species, which favor mountain and alpine conditions, with the greatest concentration in the northern hemisphere. Gentians from the Alps and Himalayas are widely admired for their beauty. Several are grown in rock gardens and other specialty gardens. A few gentians have medicinal properties. Most gentians bloom late in the season, sometimes near the end of summer, in the high mountains.

## CALIFORNIA GENERA AND SPECIES

The region has seven native genera.

### GENERA WITH SPREADING PETALS AND STAR-SHAPED FLOWERS

*Centaurium* (canchalaguas) are annuals in foothill grasslands with vivid pink flowers and oddly twisted anthers.

*Swertia* are perennials with 4 or 5 green, white, or pale purple petals often decorated with spots; each petal has a raised nectar gland.

*S. radiatum* (green gentian, deer-tongue, fig. a) is a robust mountain meadow plant with large, oval, basal leaves before flowering. At flowering, an immense panicle up to 5 feet high carries dozens of green-petalled flowers.

### GENERA WITH CUP-SHAPED OR TUBULAR COROLLAS

*Gentiana* (gentian) has appendages between the petal lobes.

*G. calycosa* (wanderer's gentian) is a clump-forming perennial with single, clear blue flowers from the high mountains.

*G. newberryi* (alpine gentian; see fig. b) has rosettes of leaves, single pale blue to white flowers with green spots, and lives in alpine meadows.

*Gentianella* (little gentians) are annuals or short-lived perennials with no appendages between the petals. They have rather small flowers.

*G. amarella* has leafy stems and narrow, pale purple flowers.

*Gentianopsis* (fringed gentians) are annuals with petals that are toothed or fringed.

*G. holopetala* (Sierra gentian) has beautiful blue flowers and lives in mountain meadows.

### OTHER GENERA

*Cicendia quadrangularis* (vernal pool gentian) is a diminutive annual of temporarily moist places with tiny, yellow flowers and 4 petals.

### a. *Swertia radiatum* (green gentian or deer tongue)

### b. *Gentiana newberryi* (alpine gentian)

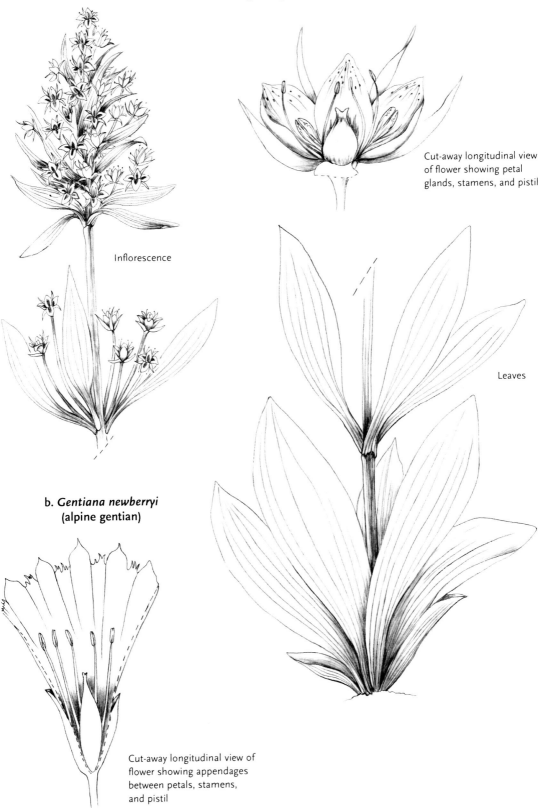

Cut-away longitudinal view of flower showing petal glands, stamens, and pistil

Inflorescence

Leaves

Cut-away longitudinal view of flower showing appendages between petals, stamens, and pistil

# GERANIACEAE (Geranium or Cranesbill Family)

RECOGNITION AT A GLANCE Herbaceous plants with round, palmately veined or pinnately compound leaves with stipules, and umbels of pink, white, or purplish flowers followed by a beaked capsule.

VEGETATIVE FEATURES Herbaceous annuals and perennials with alternate or basal round and palmately veined or pinnately compound leaves and stipules.

FLOWERS Small to medium-sized, flat, regular to irregular; white, pink, or purple; and in umbels or flat-topped cymes.

FLOWER PARTS 5 separate sepals, 5 separate petals, 5 or 10 stamens, and a single pistil with a superior, 5-lobed ovary and 5 styles joined to form a beak.

FRUITS Explosive capsules with attached styles that sometimes serve to adhere fruit sections to clothing.

RELATED OR SIMILAR-LOOKING FAMILIES The flower design of the geranium family is similar to several related families including the flaxes (Linaceae), oxalises (Oxalidaceae), and meadowfoams (Limnanthaceae). None of those families feature beaklike styles on the fruit; flaxes have 3 to 5 separate styles on each ovary and simple leaves; oxalises have 2 sets of stamens of different lengths and usually trifoliate leaves; and meadowfoams have 5 nutlets in fruit and live in temporary wetlands. (The flax and oxalis families are not detailed in this book.)

STATISTICS 750 species of wide distribution but particularly diverse in the Mediterranean region and South Africa. The scented "geraniums" (*Pelargonium* spp.) have fragrant leaves that suggest citrus, coconut, rose, and apple; their leaves are used in perfumes and to flavor food. Many other species are garden ornamentals, especially in the genera *Geranium* and *Pelargonium*. The pelargoniums are often referred to simply as "geraniums."

## CALIFORNIA GENERA AND SPECIES

The region has three genera with several native species and several introduced species.

*Erodium* (clocks, filaree) are mostly nonnative, weedy annuals with pinnately lobed to compound leaves and small, pink-purple flowers with 5 stamens.

*E. cicutarium* (see fig.) is widespread in foothill grasslands and low deserts.

*E. brachycarpum* is a common nonnative in southern California.

*Geranium* (wild geraniums; cranesbills) are nonnative and native annuals and perennials with palmately lobed leaves and 10 stamens.

*G. dissectum* is an annual weed with deeply lobed leaves.

*G. molle* is an annual weed with shallowly lobed leaves.

*G. carolinianum* is a native annual from wooded foothills.

*G. richardsonii* is a native perennial from mountain meadows with large, whitish flowers.

*Pelargonium* (garden "geraniums") are perennials and subshrubs from South Africa with round, palmately lobed and veined leaves, and irregular flowers. A few species have naturalized along the coast.

### *Erodium cicutarium* (filaree or clocks)

Upper leaf, umbel of flowers, and fruits

Flower detail showing petals, sterile stamens, and fertile stamens

# GROSSULARIACEAE (Gooseberry Family)

RECOGNITION AT A GLANCE Deciduous (mostly) shrubs with palmately lobed leaves, flowers with colored sepals and petals attached to a hypanthium, and a berry in fruit.

VEGETATIVE FEATURES Deciduous, spiny or nonspiny shrubs (one evergreen exception) with alternate, palmately lobed and veined leaves. Leaves are often fragrant.

FLOWERS Small, brightly colored, cup- or bell-shaped, and in trusslike racemes or small clusters in leaf axils.

FLOWER PARTS 4 or 5 colorful sepals, 4 or 5 smaller, cuplike petals, 4 or 5 stamens, all attached to a tubular hypanthium, and a single pistil with an inferior ovary.

FRUITS Smooth, sticky, or spiny berries.

RELATED OR SIMILAR-LOOKING FAMILIES Although many botanists placed this family with the saxifrages (Saxifragaceae), all now agree that the two families are quite different. The most commonly shared feature is the hypanthium, but otherwise gooseberry flowers have colored sepals and a clearly inferior ovary.

STATISTICS 120 species widely distributed across the northern hemisphere. A few species of gooseberry and currants (*Ribes* spp.) are cultivated for fruit and often used in preserves or dried. Several species are garden ornamentals grown for their beautiful flowers rather than their fruits. The genus serves as the alternate host for the harmful white pine blister rust, which has caused economic loss to the timber industry.

## CALIFORNIA GENERA AND SPECIES

The family has the single genus *Ribes*. The region has 31 native species.

### SPECIES WITH FLOWERS IN RACEMES AND SPINELESS STEMS

These species are currants.

*Ribes sanguineum* and varieties (pink- and red-flowering currants) have blood red to pale pink flowers and live in coastal and northern forests.

*R. malvaceum* (chaparral currant) has pale leaves, pink-purple flowers, and lives in oak woodlands and chaparral.

*R. aureum* (golden currant) has small, glossy leaves, bright yellow flowers, and red fruits, and is found in inland canyons.

*R. cereum* (wax currant) has pale leaves, pale pink to whitish flowers, and lives in the high mountains.

*R. viburnifolium* (Catalina perfume) is a sprawling shrub with lustrous, evergreen leaves and tiny dark red flowers. It comes from California's Channel Islands.

### SPECIES WITH AXILLARY FLOWERS AND TRIPLETS OF SPINES AT THE NODES

These species are gooseberries.

*Ribes menziesii* (canyon gooseberry) has nodal and internodal spines and small, dark red and white flowers. It lives in foothill forests.

*R. roezlii* (Sierra gooseberry) has nodal spines, slightly larger red or red and pink flowers, and occurs in mountain forests.

*R. montigenum* (mountain gooseberry) has nodal spines and tiny, saucer-shaped orange flowers; it lives in wet mountain forests.

*R. speciosum* (fuchsia-flowered gooseberry; see fig.) has nodal and internodal spines, narrow, tubular, bright red flowers. It is found in woodlands in south central and southern California.

### *Ribes speciosum* (fuchsia-flowered gooseberry)

Leaves and flowers

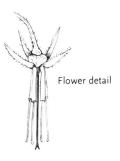

Flower detail

# HIPPOCASTANACEAE (HORSE-CHESTNUT FAMILY)

Hippocastanaceae is now included by some in Sapindaceae, the soapberry family.

RECOGNITION AT A GLANCE Small, deciduous trees with palmately compound leaves and candles of white or pale pink flowers.

VEGETATIVE FEATURES Small, deciduous trees with whitish bark and pairs of palmately compound leaves.

FLOWERS Small, unisexual or bisexual, white or pink, slightly irregular, and in dense, spikelike candles.

FLOWER PARTS 5 slightly fused, irregular sepals, 4 or 5 irregular, clawed petals, a variable number of stamens, and a single pistil with a superior, 3-chambered ovary (2 chambers usually abort).

FRUITS Leathery, globe- or pear-shaped capsules with (usually) a single, functional, poisonous, chestnutlike seed.

RELATED OR SIMILAR-LOOKING FAMILIES The horse-chestnuts have distinctive foliage—pairs of palmately compound leaves—not seen in other native plant families, and so should be easily recognized. Recent research indicates, however, that this small family is part of the mainly tropical soapberry family (Sapindaceae), which now also includes the maple family (Aceraceae).

STATISTICS 18 species in the northern hemisphere, several from eastern U.S. hardwood forests and several in Europe. The European horse-chestnuts *Aesculus hippocastanum* and the hybrid *A.* × *carnea* are widely planted trees. Although seeds resemble the edible chestnut (*Castanea* spp.), they are seriously toxic and should never be eaten.

## CALIFORNIA GENERA AND SPECIES

The region has one native species: *Aesculus californica* (California buckeye; see fig.) is a small tree with a rounded crown; white bark; fragrant, white to pale pink flowers; and leathery seed pods. It is common throughout the foothills around the Central Valley. In some places, it also approaches the coast.

The small, shrubby *A. parryi* is found in northern Baja California on the southern edge of the California floristic province.

## *Aesculus californica* (California buckeye)

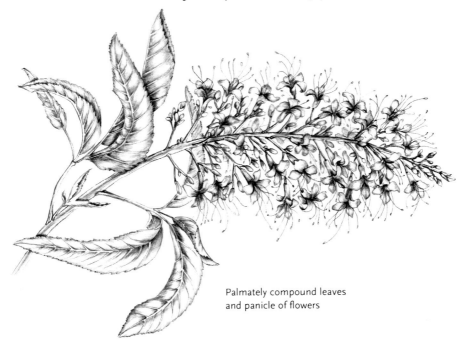

Palmately compound leaves and panicle of flowers

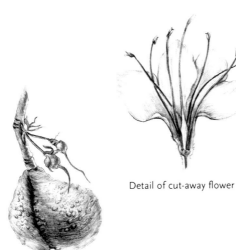

Mature fruit (and 2 aborted fruits)

Detail of cut-away flower

Seed

# HYDROPHYLLACEAE (Waterleaf Family)

Hydrophyllaceae is now considered by many to be part of Boraginaceae, the larger borage family.

RECOGNITION AT A GLANCE Mostly herbaceous plants with flower buds in fiddleheadlike coils, flowers with a forked style, and a capsule-type fruit.

VEGETATIVE FEATURES Herbaceous annuals and perennials and small shrubs with alternate (usually), simple to compound leaves.

FLOWERS Small to medium-sized; showy; cup-shaped, bell-shaped, or tubular; white, blue, purple, or yellow; and in coiled scorpioid cymes.

FLOWER PARTS 5 partly fused sepals, 5 petals joined to form a tube, 5 stamens attached to the tube, and a single pistil with a superior, 2-chambered ovary.

FRUITS Capsules with several to many seeds.

RELATED OR SIMILAR-LOOKING FAMILIES The waterleafs look a great deal like the borages (Boraginaceae) including the flower design, number of flower parts, and arrangement of flowers in scorpioid cymes. The main differences are the 2-forked style (single in borages), 2-chambered ovary (4-lobed in borages), and capsule-type fruit (nutlets in borages). Recent studies support the lumping of the waterleaf family with the borages.

STATISTICS 300 species widely distributed but with the majority found in the Americas. The greatest diversity is in western North America and the highlands of Central and South America. A few species are grown in gardens, particularly some of the native phacelias and baby-blue-eyes (*Nemophila menziesii*). Yerba santa (*Eriodictyon* spp.) has been used locally for medicinal purposes. The hairs of some species cause contact dermatitis.

## CALIFORNIA GENERA AND SPECIES

The region has 13 native genera and many species.

### WOODY-BASED GENERA

*Eriodictyon* (yerba santas; see fig. a) are small shrubs with fragrant leaves and tubular, white to purple flowers.

*Turricula parryi* (poodle dog plant) is a woody perennial with very sticky leaves and tubular purple flowers. It is abundant on burned areas in southern California.

### HERBACEOUS GENERA

FLOWERS SELDOM IN CONSPICUOUSLY COILED CLUSTERS *Hesperochiron* (meadow beauties) are perennials with rosettes of leaves and a few cup-shaped, white to pinkish flowers. They live in mountain meadows.

*Nemophila* (woodland lovers) are annuals with pinnately lobed leaves and blue or white flowers. Between their sepals are tiny, turned-down appendages (auricles).

*N. menziesii* (baby-blue-eyes) has showy blue flowers and is common in foothill grasslands and woodlands.

*N. maculata* (five-spot) has similar white flowers, each petal with a triangular purple blotch at the tip. It occurs in the Sierra foothills.

*Pholistoma* (fiesta flowers) are annuals with similar leaves and flowers but with recurved prickles by which the plants cling to their neighbors.

*P. auritum* has showy blue flowers and lives in woodlands in central and southern California.

*Eucrypta* are annuals with tiny white flowers, fragrant leaves, and no auricles between the sepals. They are common in chaparral and woodlands after burns.

GENERA WITH CLEARLY COILED FLOWER CLUSTERS *Hydrophyllum* (waterleafs) are rhizomatous perennials with headlike clusters of white, purple, or blue-purple flowers.

*H. tenuipes* is a ground cover in northern redwood forests, with dingy white flowers.

*H. occidentale* (western waterleaf) has yellowish spots on the leaves like water drops, and pale purple flowers. It is common in middle-elevation forests.

*Draperia systyla* is a wandering, rhizomatous perennial with opposite leaves and white to pale purple flowers from mid-elevation conifer forests.

*Emmenanthe penduliflora* (whispering bells) is a fire-following annual with pale yellow, bell-shaped flowers, fragrant foliage, and petals that stay on flowers in fruit. It is widespread in foothills and deserts.

*Phacelia* (see fig. b) is a large genus of annuals and perennials with simple to pinnately compound leaves. Many are difficult to key without the fruits and seeds.

*Romanzoffia* (mist-maidens) are tuberous creeping perennials with scalloped, round leaves and small, bell-shaped white flowers. They live on mossy rocks in northern forests.

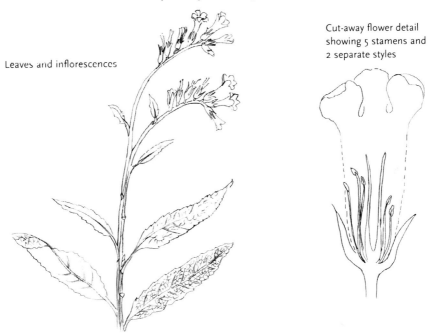

a. *Eriodictyon californicum* (yerba santa)

Leaves and inflorescences

Cut-away flower detail showing 5 stamens and 2 separate styles

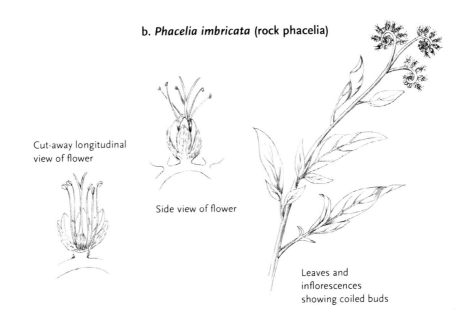

b. *Phacelia imbricata* (rock phacelia)

Cut-away longitudinal view of flower

Side view of flower

Leaves and inflorescences showing coiled buds

# IRIDACEAE (Iris Family)

RECOGNITION AT A GLANCE Perennials with sword-shaped leaves in a fanlike arrangement, and showy flowers with colored sepals and petals that emerge from a pair of leaflike bracts.

VEGETATIVE FEATURES Rhizomatous perennials with parallel-veined, sword-shaped leaves arranged in fanlike, flattened sprays that overlap at the base (equitant).

FLOWERS Showy, regular, flaglike, saucer- or funnel-shaped, and in small clusters protected in bud by 2 leaflike bracts.

FLOWER PARTS 3 colorful sepals alternate with 3 similar or differently shaped petals; 3 stamens, all usually attached to a hypanthium; and a single pistil with an inferior, 3-chambered ovary.

FRUITS Capsules with many seeds.

RELATED OR SIMILAR-LOOKING FAMILIES Many species have flowers that resemble the designs in the lily family (Liliaceae), but they differ by having 3 stamens (most lilies have 6) and an inferior ovary (most lilies have superior ovaries). In case of doubt, the pairs of leaflike bracts that protect the flower buds are diagnostic. The iris family is usually recognized by its special leaf design and arrangement but *Tofieldia* and *Narthecium* in the lily family also have equitant leaves, and several species of rush (*Juncus* spp.) have them as well. In the case of the rushes, the tiny flowers are wind-pollinated and have bronze or greenish tepals. Equitant leaves also appear in such nonnative families as the bloodroots (Haemadoraceae).

STATISTICS 1,500 species widely distributed through the world and especially diverse in the Cape Province region in South Africa. A few species are used medicinally but the most important use is as garden ornamentals, including the many species of iris (*Iris* spp.) and such South African "bulbs" as *Gladiolus* spp., *Watsonia* spp., *Crocosmia* (montbretia), *Babiana* spp., *Ixia* spp., and *Freesia* spp.

## CALIFORNIA GENERA AND SPECIES

The region has two native and nine nonnative genera.

NATIVE GENERA
*Iris* (wild irises; flags) have large flowers with 3-colored, turned-down sepals (falls) alternating with 3 upright petals (standards). In addition, there are 3 petallike styles in the center of the flower, which hide the 3 stamens.

SPECIES WITH NO APPARENT HYPANTHIUM *Iris missouriensis* (Sierra or mountain iris) has stiff, bluish green leaves and pale blue flowers, and lives in wet mountain meadows.

*I. longipetala* (coast iris) has similar leaves and flowers and grows in moist grasslands near the Bay Area.

SPECIES WITH WELL DEVELOPED HYPANTHIUMS *Iris douglasiana* (Douglas iris; see fig. a) is widespread along the coast; it features glossy, dark green leaves, and flowers in blue, purple, lavender, white, and pale yellow. It is often hybridized with other species to create garden cultivars known as Pacific Coast irises.

*I. macrosiphon* (ground iris) occurs in the north Coast Ranges and northern Sierra foothills. It has narrow, pale green leaves and yellow, blue, or purple flowers borne close to the ground.

*I. hartwegii* (rainbow iris) occurs in the yellow pine belt of the Sierra and other high mountains, and has narrow, bluish green leaves and pale yellow flowers carried above the leaves.

*Sisyrinchium* has small perennials with umbel-like clusters of saucer-shaped blue, purple, or yellow flowers:

*S. bellum* (blue-eyed grass; see fig. b) features blue-purple flowers with a yellow center and is widespread in the foothills.

*S. californicum* (yellow-eyed grass) has yellow flowers and is limited to seeps along the coast.

NONNATIVE GENERA
The following three nonnative genera of the region all hail from South Africa.

*Watsonia* has tall leaves and spikes of white, pink, or orange flowers. It is naturalized along the north coast.

*Freesia retrofracta* has shorter leaves and spikes of fragrant, white flowers. It is occasionally naturalized.

*Crocosmia* (montbretia) has stoloniferous corms and spikes of bright orange flowers. It has invaded coastal forests.

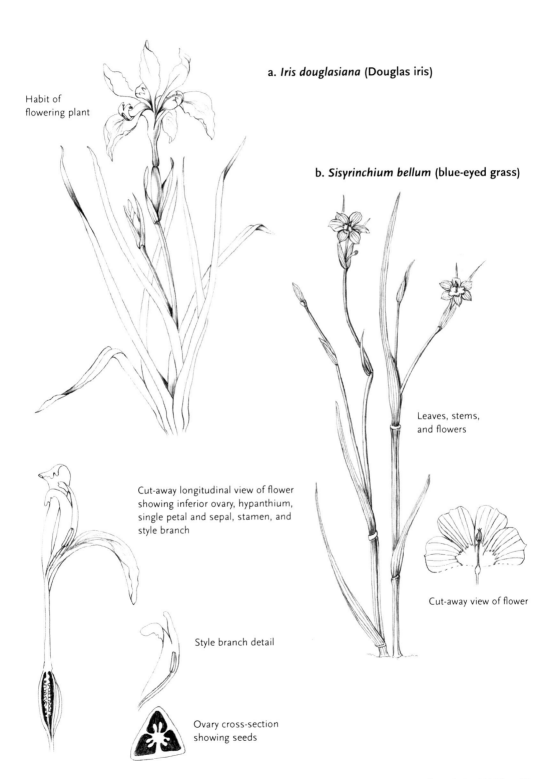

a. *Iris douglasiana* (Douglas iris)

b. *Sisyrinchium bellum* (blue-eyed grass)

# JUNCACEAE (Rush Family)

**RECOGNITION AT A GLANCE** Grasslike plants with pithy green stems (not hollow as in grasses); tiny flowers with 6 green, white, or brown tepals; and a capsule-type fruit.

**VEGETATIVE FEATURES** Herbaceous annuals and perennials usually with short to long rhizomes, fibrous roots, pithy stems, and one of several leaf designs. Some have rounded green stems with scalelike leaves; others have grasslike or flat, equitant, irislike leaves.

**FLOWERS** Tiny, bisexual, star-shaped, greenish or brownish, and in umbels, heads, or cymes, sometimes with short bracts.

**FLOWER PARTS** 6 tepals, 3 or 6 stamens, and a single pistil with a 1- or 3-chambered, superior ovary and feathery stigmas.

**FRUITS** Capsules.

**RELATED OR SIMILAR-LOOKING FAMILIES** Although rushes join the ranks of grasslike plants that also include the grass family (Poaceae) and sedge family (Cyperaceae), they are generally easy to tell apart by the combination of leaf designs, pithy stems, flowers with tepals, and capsules in fruit. Grasses feature hollow stems and 2 ranks of leaves, their flowers lack tepals, and their fruits are grainlike caryopses. Sedges have solid, triangular stems, 3 ranks of leaves, flowers with bristles, and their fruits are achenes.

**STATISTICS** About 325 species well distributed across the temperate world, chiefly in wet habitats. Besides local uses of rushes for mats and such, the family has no economic value.

## CALIFORNIA GENERA AND SPECIES

The region has two native genera with many species.

*Juncus* (rushes; see figs. a and b) have hairless leaves in a variety of designs and usually live where the water table is high. The several species fall naturally into four groups: small annuals; perennials with cylinder-shaped leaves; perennials with flattened, irislike leaves; and perennials with scalelike leaves and stiff, green stems ending in a cluster of flowers and a pointed, leaflike bract.

*Luzula* (woodrushes) have soft, flat, leaves with crimped or kinked hairs, and live in dry woods.

### a. *Juncus* sp. (rush)

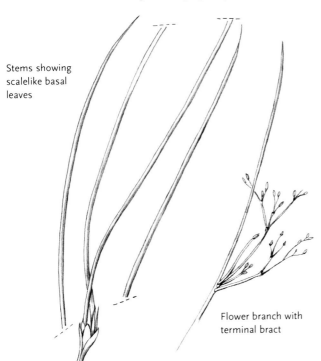

Stems showing scalelike basal leaves

Flower branch with terminal bract

Cut-away longitudinal view of flower with tepals, stamens, and pistil

### b. *Juncus bufonius* (toad rush)

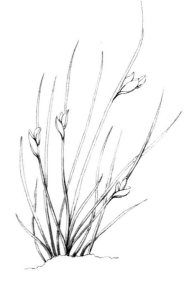

Habit of plant

# LAMIACEAE (Mint Family)

**RECOGNITION AT A GLANCE** A combination of square stems, opposite leaves, and strong fragrance usually suffices to identify the family. In addition, the 2-lipped flowers have 4-lobed ovaries.

**VEGETATIVE FEATURES** Herbaceous annuals, perennials, and shrubs with square stems on the new growth; and opposite, often simple, usually highly fragrant leaves.

**FLOWERS** Showy, sometimes small, generally irregular and 2-lipped, and arranged in heads or whorled spikes.

**FLOWER PARTS** Variable number (often 5) of fused sepals, 5 irregular, usually 2-lipped petals joined to form a tube, 2 or 4 stamens joined to the tube, and a single pistil with a 4-lobed, superior ovary.

**FRUITS** 4 one-seeded nutlets per flower.

**RELATED OR SIMILAR-LOOKING FAMILIES** Several other families feature flowers with a 2-lipped design and similar number of stamens: Scrophulariaceae (figwort family), Acanthaceae (acanth family), Bignoniaceae (trumpet vine family), Orobanchaceae (broomrape family), and others. None of those families has a 4-lobed ovary or highly fragrant leaves. The mints are also very closely related to the verbena family (Verbenaceae), which differs chiefly in having slightly irregular flowers, less-fragrant leaves, sometimes woody stems, and ovaries that are not as deeply 4-lobed. (The broomrape, acanth, and trumpet vine families are not detailed in this book.)

**STATISTICS** 5,500 species widely distributed across the world. Many live in dry habitats. The family is richly represented in Mexico, South America, the Mediterranean region, and California. The fragrant, volatile oils in the foliage are used medicinally and to flavor food. The many culinary herbs include rosemary, thyme, oregano, sage, lavender, basil, and mint. Many beautiful ornamentals belong to this family including the currently popular salvias.

## CALIFORNIA GENERA AND SPECIES

The region has 27 genera, seven of which are completely nonnative.

*NATIVE GENERA*

**SHRUBBY GENERA** *Trichostema lanatum* and *T. parishii* (woolly blue-curls) are small shrubs with narrow, bright green leaves and whorled spikes of woolly-sepaled, purple and blue flowers with curled stamens. (Several other species are small summer annuals with vinegar-scented leaves.)

*Lepechinia* (pitcher-sages) are small shrubs with wrinkled, highly scented, triangular leaves; nodding, purple or white flowers; and inflated, pitcher-shaped sepals in fruit.

*Salvia* (sages) are mostly small, semievergreen shrubs with varied, sage-scented leaves; whorled spikes of blue, purple, pink, or white flowers; and two stamens. *S. mellifera* (black sage; see fig. a), *S. leucophylla* (purple sage), and *S. apiana* (white sage) are abundant in southern California's coastal sage scrub and chaparral.

*S. spathacea* (hummingbird sage) is an herbaceous perennial with bold, whorled spikes of rose-pink flowers.

*S. columbariae* (chia) and *S. carduacea* (thistle sage) are dryland annuals with spiny bracts and blue flowers (chia) or lavender-purple flowers (thistle sage).

**HERBACEOUS OR WOODY-BASED PERENNIAL GENERA** *Agastache* (horsemints) are tall mountain-meadow perennials with small purple flowers in whorled spikes.

*Satureja* (yerba buena, native savories) are fragrant ground covers or small, bushy plants with white flowers (*S. mimuloides* has orange flowers).

*Stachys* (woodmints or hedge-nettles; see fig. b) are rhizomatous perennials with wrinkled, unpleasantly scented leaves, and whorled spikes of purple, pink, red, or white flowers. Most live in woodlands or along streams.

*Scutellaria* (skullcaps) are clump-forming perennials (sometimes with tuberous roots) with small, unscented leaves, and axillary whorls of white, blue, or purple snapdragonlike flowers.

*Monardella* (coyote-mints or western pennyroyals) are bushy, woody-based perennials with highly fragrant leaves and headlike clusters of tubular flowers.

*M. villosa* (common coyote-mint; see fig. c) has blue or purple flowers and lives throughout the foothills.

*M. odoratissima* (mountain coyote-mint) has similar flowers but lives on rocky slopes in the high mountains.

*M. macrantha* (hummingbird mint) is a creeping plant with very long, tubular, red or orange flowers from the central and south Coast Ranges.

*Pogogyne* (vernal pool mints) are tiny annuals with highly fragrant leaves and axillary whorls of pink or purple flowers; they live by vernal pools.

*Acanthomintha* (thorn-mints) are annuals with spiked whorls of white or purple flowers with highly spiny bracts. They are uncommon on dry, rocky slopes.

### NONNATIVE, OFTEN-WEEDY GENERA

*Mentha* (mints) are invasive, rhizomatous perennials with whorls of white or purple, slightly irregular flowers.

*M. spicata* (spearmint) and *M. pulegium* (pennyroyal) are common in moist areas. (One species—*M. arvensis*—is native.)

*Prunella vulgaris* (self-heal) is a ground-hugging perennial with nearly odorless leaves and whorled spikes of deep blue flowers. Variety *vulgaris* is introduced, var. *lanceolata* is native and lives in coastal areas.

*Marrubium vulgare* (horehound) is an annual in dry places; it has wrinkled, whitish leaves, and whorled spikes of small white flowers.

*Lamium amplexicaule* (henbane) is an annual garden weed with floppy stems, oval leaves, and axillary whorls of pink-purple flowers.

*Melissa officinalis* (lemon balm) is a vigorous perennial with bright green, lemon-scented, wrinkled leaves and whorled spikes of white flowers.

*Glecoma hederacea* (ground-ivy) is a creeping ground cover from coastal forests.

# LAMIACEAE (Mint Family)

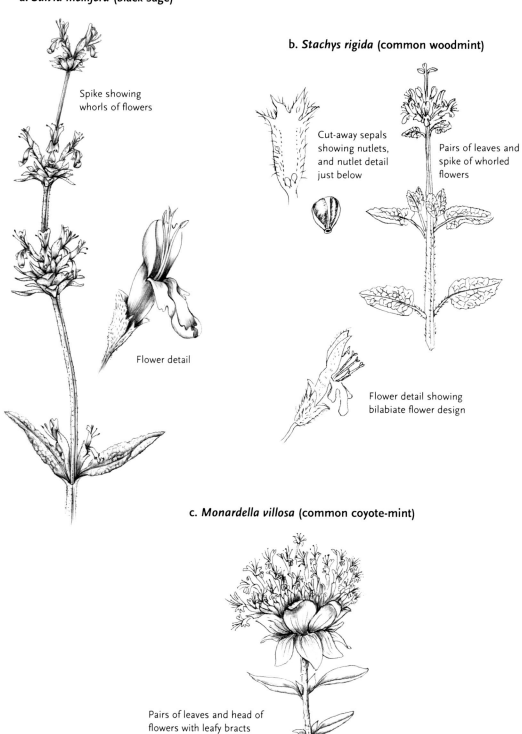

a. *Salvia mellifera* (black sage) — Spike showing whorls of flowers; Flower detail

b. *Stachys rigida* (common woodmint) — Cut-away sepals showing nutlets, and nutlet detail just below; Pairs of leaves and spike of whorled flowers; Flower detail showing bilabiate flower design

c. *Monardella villosa* (common coyote-mint) — Pairs of leaves and head of flowers with leafy bracts

# LAURACEAE (Laurel Family)

RECOGNITION AT A GLANCE Evergreen trees with simple, strongly aromatic leaves; small yellow or whitish flowers; and avocadolike drupes.

VEGETATIVE FEATURES Evergreen (shrubs) or trees with highly fragrant simple, alternate, entire, leaves.

FLOWERS Small, unisexual or bisexual, yellow or whitish, saucer-shaped, and often in umbel-like clusters.

FLOWER PARTS 2 whorls of 3 colored tepals, 3 rows of 3 stamens—each stamen opening by a lidlike flap—and a single pistil with a superior ovary.

FRUITS 1-seeded, fleshy, avocadolike drupes.

RELATED OR SIMILAR-LOOKING FAMILIES The distinctive fruit and aromatic leaves make this family easy to recognize. The laurel family is recognized for its position among primitive flowering plant families, and is now separated into an early group known as the basal dicots. The laurel family is among a handful of native dicot families with flower parts in threes that also includes the birthwort family (Aristolochiaceae) and barberry family (Berberidaceae). Both of those families are easily told apart: birthworts are herbaceous plants with heart-shaped leaves and oddly shaped and colored flowers; barberries have compound leaves and yellow or white flowers with 3 rows of tepals.

STATISTICS 2,200 species, most diverse in the tropics. A few live in temperate forests. Plants of commercial value include avocado (*Persea americana*), cinnamon (*Cinnamomum zeylanicum*), sassafras (*Sassafras albida*), and the camphor tree (*Cinnamomum camphoratum*), widely planted as a street tree in California. Bay-laurel (*Laurus nobilis*) is a classic Mediterranean plant whose leaves flavor food.

## CALIFORNIA GENERA AND SPECIES

The region has a single native species.

*Umbellularia californica* (California bay, bay-laurel, pepperwood, myrtlewood; see fig.) is a shrub (on serpentine soils) or a large, multi-trunked tree with lance-shaped leaves, yellow flowers that appear in winter, and purple drupes. It lives in many habitats, including coastal forests and the yellow pine belt in the mountains.

# LAURACEAE (Laurel Family)

## *Umbellularia californica* (California bay or pepperwood)

Leafy branch with flowers

Fruit

Flower detail showing petals and several stamens

# LILIACEAE (Lily Family)

IMPORTANT NOTE: The lily family has been controversial for many years, and different books have treated it in many different manners. Now with accumulating evidence from many lines of study, we can no longer accept the lily family the way it has sometimes been recognized, as an all-encompassing group held together by a similar basic flower plan. It is beyond the scope of this book to detail all of the proposed changes at hand, but by some estimates California now has 15 splinter families. Many of these families are no longer considered remotely related, and they undoubtedly represent several distinct lines of evolution.

For convenience, I am placing many of the members of this group in this extended version of the lily family although I have removed three important groups: the agave family (Agavaceae), onion family (Alliaceae), and brodiaea family (Themidaceae), all of which are easily recognized as distinct families without any contradictions. In discussing genera and species, I have also remarked on the new family assignment of many of the genera described below.

As a result, my coverage of the lily family excludes all large, woody-based plants with massive clusters of flowers, and most bulb- and corm-bearing species that display their flowers in umbels.

RECOGNITION AT A GLANCE Often bulb- or corm-bearing plants with simple, parallel-veined leaves and panicles or racemes of showy flowers with (usually) 6 colored tepals and stamens.

VEGETATIVE FEATURES Mostly geophytes (plants from corms, tubers, rhizomes, or bulbs) with simple, (usually) untoothed leaves and (usually) parallel veins.

FLOWERS Small to showy, of many different colors and shapes, regular; borne singly or arranged in racemes or panicles.

FLOWER PARTS (Usually) 6 colored tepals (similar sepals and petals), 6 (occasionally 3) stamens, and a single pistil with a superior, 3-chambered ovary.

FRUITS Capsules with many seeds or fleshy berries.

RELATED OR SIMILAR-LOOKING FAMILIES The other monocot family (besides the lily splinter families, see above) sometimes confused with the lilies is the distantly related iris family (Iridaceae), which differs by having sword-shaped, equitant leaves, 3 stamens (most lilies have 6), an inferior ovary (lilies have superior ovaries), and flower buds hidden inside a pair of leaflike bracts. As mentioned above, the Liliaceae is controversial, and all recent botanists have split it into many separate families based on major differences in embryo development, seed details, leaf-vein patterns, and DNA studies. I have treated three lilylike families elsewhere in this book: the Agavaceae (agave family) has large, fibrous leaves attached to a woody base and immense inflorescences of flowers; the Alliaceae (onion family) has an onion odor and bracted umbels of flowers; the Themidaceae (brodiaea family) also has bracted umbels of flowers but lacks any onion odor.

STATISTICS Currently, the much diminished lily family has around 1,000 species worldwide and is poorly represented in tropical rainforests. (Several live in tropical highlands.) The family is particularly diverse in South Africa, the Mediterranean Basin, Australia, and California. Edible plants include asparagus (*Asparagus officinalis*), actually, now in the Asparagaceae; many of our native bulbous lily relatives also served as food for the California Indians. *Aloe vera* (now Asphodelaceae) is famed for the healing gel inside its fleshy leaves, which is used in many herbal remedies. Numerous garden ornamentals, particularly favorite bulbs, belong to this family, including lilies (*Lilium* spp.), day-lilies (*Hemerocallis* spp.), grape-hyacinths (*Muscari* spp.), hyacinths (*Hyacinthus* spp.), and tulips (*Tulipa* spp.).

## CALIFORNIA GENERA AND SPECIES

The lily family is one of the most varied and characteristic elements of the flora. There are many genera, few of which are entirely non-native.

### GENERA WITH STEM LEAVES ONLY

*Trillium* (trilliums) have a single whorl of three broad, net-veined leaves, and a single red, pink, yellowish, or white flower with 3 green sepals and 3 colored petals. They live in wooded habitats. Currently trilliums are placed in their own small family Trilliaceae, characterized by net-veined leaves and flowers with green sepals and colored petals.

*Smilacina* (false Solomon's seals; see fig. a) have unbranched stems with several ovate leaves and racemes or panicles of tiny, white, star-shaped flowers. They also live in wooded habitats.

*Disporum* (fairy bells) have branched stems with several clasping, ovate leaves, and hanging, bell-shaped, white or greenish flowers. They live in wooded habitats.

*Streptopus amplexifolius* (twisted stalk) has similar stems and leaves to fairy bells but the hanging flowers are on tiny stalks that have a sharp, kneelike bend.

*Smilax californica* (green-brier) is a semi-woody, climbing vine with weak prickles, broad leaves, and tiny greenish flowers. Currently green-brier belongs to its own family Smilacaceae, a family noted for woody, climbing or shrubby prickly stems and umbel-like clusters of greenish flowers.

*Lilium* (true lilies; see fig. b) have whorls or whorllike clusters of narrow leaves up single stems, and horizontal or nodding bell-shaped flowers (or flowers with tepals curled back) and versatile anthers (the anthers pivot around their point of attachment to the filament). Many species have orange flowers with dark spots.

*Fritillaria* (fritillaries; checker-lilies, mission bells) have similar leaves to lilies but the bell-shaped flowers are often checkered or mottled with brown or purple, and the anthers are fixed in position on the filaments.

*Veratrum* (corn-lily) have stout stalks with broad, pleated leaves and large panicles of star-shaped white (usually) flowers. They live mostly in wet meadows or bogs.

### GENERA WITH MOSTLY BASAL LEAVES

*Scoliopus bigelovii* (slink pods; fetid adder's tongue) has pairs or triplets of broad leaves that are darkly mottled, and clusters of dark purple flowers with an acrid odor. It lives in moist coastal conifer forests and blooms in winter.

*Clintonia* (bead-lilies) have rosettes or small clusters of broad, oval, shiny leaves and panicles or single bell-shaped, pink-purple (*C. andrewsiana*) or white flowers (*C. uniflora*). They live in conifer woods.

*Erythronium* (fawn- or glacier-lilies) have 2 or 3 broad, mottled or solid green leaves and one to several large, nodding white, yellow, or pinkish flowers with recurved petals.

*Hastingsia alba* has numerous, grasslike leaves and racemes or panicles of tiny white flowers. It lives in boggy places in the north.

*Stenanthium occidentale* (bronze bells) has several narrow, grasslike leaves and racemes of hanging, bell-shaped bronze and greenish flowers. It lives in northwestern mountains.

*Narthecium californicum* has green, irislike leaves and racemes of bright yellow flowers.

*Tofieldia glutinosa* (bog asphodel) has bluish green, irislike leaves and narrow, sticky racemes of small white flowers. These last two genera live in bogs in the northern part of the state. Both *Narthecium* and *Tofieldia* are considered unrelated to most other lily relatives, and many botanists place each in its own respective family, Nartheciaceae and Tofieldiaceae.

*Camassia quamash* (camas) has grasslike leaves and racemes of showy, star-shaped, clear blue flowers. It lives in wet meadows.

*Zigadenus* (death-camases, starlilies) have grasslike leaves and racemes or panicles of star-shaped, cream-colored flowers, each tepal with a yoke-shaped yellowish gland. Their bulbs are deadly poisonous, whereas camas bulbs are considered highly edible.

*Chlorogalum* (soap plants, amoles) have grasslike, sometimes wavy leaves; very large, soapy bulbs; and panicles of small, wide open, white or purple flowers. They are abundant in grasslands and woodlands.

*Xerophyllum tenax* (bear-grass) has dense clumps of numerous, grasslike, tough leaves and very tall, dense panicles of numerous, tiny, star-shaped, white flowers. The leaves were used in Indian basketry.

*Calochortus* (various common names) often have a single large, long, strap-shaped basal leaf; several leaflike bracts; and large, showy, bell-, saucer-, tulip-, or globe-shaped flowers of great beauty. The sepals are clearly smaller and often less colored than the petals, and the shape and color of the nectar glands are important in identifying species. This large genus is divided roughly into three groups.

Globe-tulips have hanging, closed, globe-shaped yellow, pink, or white flowers. They occur in woodlands and include *C. amabilis*, *C. pulchellus*, *C. amoenus*, and *C. albus*.

Star-tulips or pussy ears have upright, saucer-shaped, white, blue, pink, yellow, or purple flowers sometimes densely bearded with hairs. Species include *C. umbellatus*, *C. tolmiei*, *C. coeruleus*, *C. uniflorus*, and *C. monophyllus*.

Mariposa-tulips (see fig. c) have large, upright, tulip-shaped flowers in a rainbow of colors. The petals often are penciled, splotched, and striped with other colors, resembling butterfly wings. This group includes many diverse species.

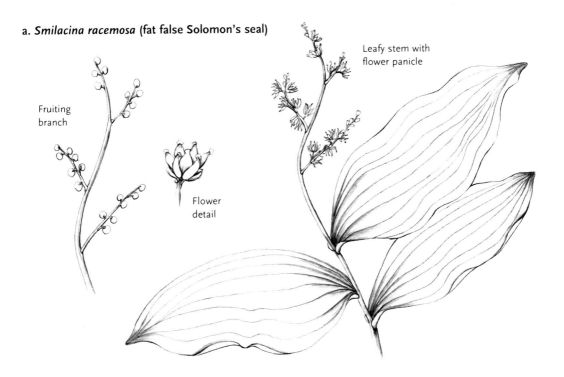

a. *Smilacina racemosa* (fat false Solomon's seal)

Fruiting branch

Flower detail

Leafy stem with flower panicle

# LILIACEAE (Lily Family)

**b. *Lilium pardalinum* (leopard lily)**

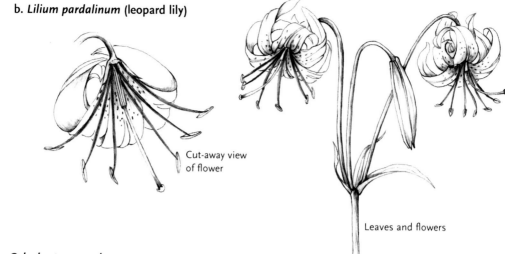

Cut-away view of flower

Leaves and flowers

**c. *Calochortus superbus* (superb mariposa-tulip)**

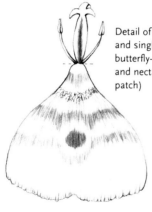

Detail of 2 stamens, pistil, and single petal showing butterfly-wing-like markings and nectar gland (the hairy patch)

# LOASACEAE (Blazing Star Family)

**RECOGNITION AT A GLANCE** Herbaceous plants with leaves rough-textured or with stinging hairs, and flowers with (usually) numerous stamens and an inferior ovary.

**VEGETATIVE FEATURES** Annuals, perennials, and subshrubs with simple, often toothed leaves covered with rough-textured or stinging hairs.

**FLOWERS** Flowers vary from small to large and showy; are star-, vase- or cup-shaped; and are yellow, cream-colored, or orange.

**FLOWER PARTS** 5 separate sepals, 5 petals sometimes lightly joined, (usually) numerous stamens, and a single pistil with an inferior ovary.

**FRUITS** Capsules with (usually) numerous seeds.

**RELATED OR SIMILAR-LOOKING FAMILIES** Few families are likely to be confused with the blazing stars although some showy species of *Mentzelia* have flowers of similar color and shape to members of the evening-primrose family (Onagraceae). That family, however, has 4 sepals and petals, a fixed number of stamens, and (usually) a hypanthium.

**STATISTICS** 200 species found in warm, dry, or subtropical habitats. Although species are widely distributed, many species occur in the New World and are richly represented in western North America. Only a few are grown in gardens.

CALIFORNIA GENERA AND SPECIES

The region has three native genera and 33 species.

*STAMENS ONLY 5 AND OVARY SINGLE-SEEDED*
*Petalonyx* (sandpaper plant) has three desert species with tiny white flowers.

*P. thurberi* is the most common species and widely distributed in dry washes.

*STAMENS NUMEROUS AND OVARY MANY-SEEDED*
*Mentzelia* (blazing star) has separate petals and roughened leaves without stinging hairs. The 20-some species range from annuals with tiny flowers to robust biennials and perennials with large, showy flowers.

*M. laevicaulis* (see fig.) is a widely branched biennial with large, pale yellow blossoms that open early in the day. It is widespread on gravelly slopes of summer-hot areas.

*M. lindleyi* is a showy annual with large, bright yellow blossoms and orange stamens. It is restricted to scree in the central Coast Ranges; the closely similar *M. crocea* is abundant in the southern Sierra.

*M. involucrata* (satin blazing star) is a tall annual with large, satiny, pale yellow petals flushed with pink; it is locally abundant in deserts in years of ample rain.

*Eucnide* (rock-nettle) has leaves with stinging, nettlelike hairs and petals joined at the base. The flowers are greenish in *E. rupestris* (rare in southernmost California) and white to pale yellow in *E. urens* (common in the Death Valley region).

***Mentzelia laevicaulis*** **(common blazing star)**

Upper stem with leafy bracts, flower buds, and open flower

# MALVACEAE (Mallow Family)

Malvaceae now includes Sterculiaceae, the cacao family.

RECOGNITION AT A GLANCE Herbs, shrubs, and small trees with broad, usually palmately lobed and veined leaves and showy flowers with (usually) numerous stamens fused into a tube by their filaments.

VEGETATIVE FEATURES Herbaceous annuals, perennials, shrubs, and small trees with usually broad, simple leaves often with palmate lobes and veins; stipules; and (often) starlike hairs (use a strong hand lens).

FLOWERS Often showy, wide open, and solitary or arranged in clusters, sometimes in leaf axils.

FLOWER PARTS 5 sepals, often sepal-like bracts below the sepals, 5 petals often lightly joined at the base to a tube formed by the fused filaments of the (usually) numerous stamens, and a single pistil with a superior, often many-segmented, cheesewheel-like ovary, and several styles.

FRUITS (Usually) one to few-seeded schizocarps or woody capsules.

RELATED OR SIMILAR-LOOKING FAMILIES The mallow family is unlikely to be confused with other families because of the (usually) numerous stamens fused into a hollow tube. The leaf design is somewhat suggestive of the geranium family (Geraniaceae). The closely related cacao family (Sterculiaceae), which features 5 stamens fused into a hollow tube, has now been combined with the Malvaceae (see below).

STATISTICS 2,000 species widely distributed throughout the world and especially diverse in dry areas and the tropics. Cotton (*Gossypium* spp.) is an important source of fibers as well as the source of cottonseed oil. Okra *(Hibiscus esculentus)* is an important vegetable used in several cuisines; other species of hibiscus have edible leaves, or flowers used in herbal teas. The original marshmallows were prepared from the mucilaginous roots of the European marshmallow. The family is noted for many handsome, sometimes tender, garden ornamentals including hibiscus (*Hibiscus rosa-sinensis*), hollyhock (*Alcea rosea*), malva rosa (*Lavatera* spp.), and flowering-maples (*Abutilon* spp.). Cacao (*Theobroma cacao*), the source of chocolate, is now also a member of this far-flung family.

CALIFORNIA GENERA AND SPECIES

The region has 14 native and four nonnative genera.

*GENERA WITH LONG, SLENDER, RIBBONLIKE STIGMAS*

*Alcea rosea* (hollyhock) is an occasional garden escape from Europe with tall spikes of pink, white, or pale yellow flowers.

*Lavatera assurgentiflora* (malva rosa) is a green-twigged evergreen shrub from California's Channel Islands with maplelike leaves and showy, rose-purple flowers.

*Malva* (mallows; cheeses) are invasive European weeds with round leaves and small, axillary, pink or purple flowers.

*Sidalcea* (checkerblooms or checkermallows) are annuals or perennials with rounded basal leaves that are shallowly lobed, stem leaves that are deeply dissected, and racemes of showy, purple to rose-pink blossoms. Several species are rare.

*S. malviflora* (see fig. a) is a common grassland and woodland perennial with pink flowers similar to those of the hollyhock.

*S. oregana* var. *spicata* (mountain hollyhock), from wet mountain meadows, has narrow spikes of deep pink to purple flowers.

*S. glaucescens* (mountain checkerbloom) grows in dry, rocky meadows and has pale purple flowers.

*GENERA WITH KNOBLIKE OR BLUNT STIGMAS*

GENERA WITHOUT OR WITH FEW BRACTLETS BELOW THE FLOWERS *Sphaeralcea* (bush or apricot-mallows) are desert subshrubs with flowers in various shades of orange and red.

*S. ambigua* (apricot-mallow) is widespread and common.

GENERA WITH 1 TO 3 BRACTLETS BELOW EACH FLOWER *Malacothamnus* (bush-mallows) are fast-growing, soft woody shrubs with dense clusters of white, pink, or purple flowers in leaf axils. They often appear in quantity after fire.

*M. fasciculatus* is the most variable and widespread species, extending from the Bay Area south through the mountains of southern California. *M. hallii* (Hall's bush mallow; see fig. b) is a rare species illustrated here but not included in the current *Jepson Manual*.

*Malvella leprosa* (alkali mallow) is an herbaceous perennial with yellow flowers and white, scurfy leaves that grows on dry, salty soils in summer-hot areas.

*Iliamna* (globe-mallows, bush mallows) are woody-based perennials from the northwestern corner of the state with globe-shaped, rose-purple flowers.

*I. bakeri* occurs on dry lava soils north of Mt. Shasta.

*I. latibracteata* is uncommon in montane conifer forests in Humboldt and Del Norte counties.

## GENERA WITH ONLY 5 STAMENS (THE FORMER STERCULIACEAE)

*Fremontodendron* (fremontia, flannel bush, two native species) are prostrate woody shrubs to small trees that have broad leaves felted with irritating hairs; large, saucer-shaped, orange to yellow flowers; and woody, 5-parted capsules. They live in the chaparral and on the edges of forests mostly in southern California, although small populations appear as far north as the central Sierra Nevada and Napa County in the Coast Ranges.

## OTHER GENERA

*Hibiscus* has woody capsules instead of schizocarps.

*H. lasiocarpus* (marsh-mallow) is a robust perennial from the Sacramento River Delta region, with large white flowers centered red.

# MALVACEAE (Mallow Family)

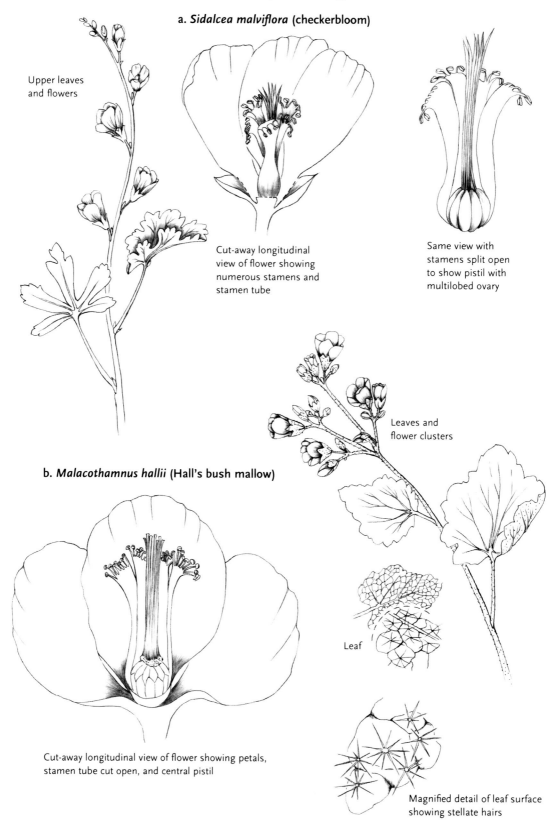

a. *Sidalcea malviflora* (checkerbloom)

Upper leaves and flowers

Cut-away longitudinal view of flower showing numerous stamens and stamen tube

Same view with stamens split open to show pistil with multilobed ovary

b. *Malacothamnus hallii* (Hall's bush mallow)

Leaves and flower clusters

Leaf

Cut-away longitudinal view of flower showing petals, stamen tube cut open, and central pistil

Magnified detail of leaf surface showing stellate hairs

## MYRTACEAE (Myrtle Family)

RECOGNITION AT A GLANCE Woody plants with simple, entire, highly fragrant leaves and flowers with numerous stamens and an inferior ovary.

VEGETATIVE FEATURES Shrubs and trees with (usually) alternate (sometimes opposite in young plants), simple, entire, highly fragrant leaves.

FLOWERS Vary from small to large and showy and are white, cream-colored, yellow, purple, pink, or red. Flowers are mostly arranged in dense spikes, clusters, or umbels.

FLOWER PARTS (In most) 4 or 5 separate sepals, (in most) 4 or 5 separate petals, many to numerous stamens, and a single pistil with an inferior ovary.

FRUITS Woody capsules or fleshy berries.

RELATED OR SIMILAR-LOOKING FAMILIES Although other woody families such as the laurel family (Lauraceae) and rue family (Rutaceae) also have simple, highly fragrant leaves, the flower details and fruits are different. Laurels feature flower parts in multiples of 3 and have drupelike fruits; rues have flowers with a fixed number of stamens and sometimes capsular or samaralike fruits.

STATISTICS 3,000 species concentrated in the tropics and also highly diversified in Australia. Cloves (*Syzygium aromaticum*) and allspice (*Pimenta dioica*) are important for flavoring foods, while various guavas and their relatives are cultivated for their edible fruits. Aromatic oils are extracted from the leaves of eucalyptuses. Many myrtles are cultivated in warm temperate and subtropical gardens including common myrtle (*Myrtus communis*), a wide variety of eucalypts (*Eucalyptus* spp.), lilly-pilly tree (*Acmena smithii*), bottlebrushes (*Callistemon* spp.), paperbarks (*Melaleuca* spp.), tea trees (*Leptospermum* spp.), and many more.

### CALIFORNIA GENERA AND SPECIES

The region has no native species, three nonnative genera, and 11 species.

*Luma apiculata* is a small tree with fleshy fruits naturalized near the coast.

*Leptospermum laevigatum* (Australian tea-tree) is a large shrub or small tree naturalized near the coast with white, 5-petaled flowers and woody capsules.

*Eucalyptus* (stringy-barks, gums, and others) are naturalized trees featuring flowers covered in bud by a cap of fused petals and sepals that drops off to display bunches of long, brightly colored stamens.

*E. globulus* (blue gum; see fig.) is the tree mostly widely naturalized; it features powdery blue juvenile leaves on saplings, bark that peels in multicolored strips, and white flowers.

*E. sideroxylon* (red ironbark) has craggy, dark red-brown bark and pink flowers.

# MYRTACEAE (Myrtle Family)

## *Eucalyptus globulus* (blue gum)

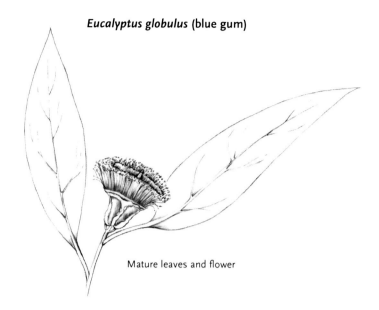

Mature leaves and flower

Juvenile leaves

Woody capsule

# NYCTAGINACEAE (Four O'Clock Family)

RECOGNITION AT A GLANCE Herbaceous plants with pairs of simple leaves, flowers with petal-like sepals (and no petals), and an apparently inferior ovary.

VEGETATIVE FEATURES Annuals, taprooted or tuberous perennials, and subshrubs with pairs of (usually) rounded, simple, entire leaves.

FLOWERS Small; often flared tubular in shape; white, pink, red, yellow, or rose-purple; and often arranged in dense clusters that are sometimes surrounded by sepal-like bracts (involucres).

FLOWER PARTS (Usually) 4 or 5 colored, petal-like sepals joined to a tubular hypanthium, no petals, 4 or 5 stamens, and a single pistil with an apparently inferior ovary. The ovary is not truly joined to the hypanthium and so is technically superior.

FRUITS 1-seeded achenes or nutlets.

RELATED OR SIMILAR-LOOKING FAMILIES The combination of opposite leaves and colored sepals atop a hypanthium tube with a seemingly inferior ovary is unique to this family and not likely to be confused with other families. Note that the sand-verbenas (*Abronia* spp.) are not related to the true verbenas (*Verbena* spp.) and differ in their flower details; true verbenas have green sepals and colored petals and feature a clearly superior ovary and no hypanthium.

STATISTICS 300 species from dry habitats as well as tropical shrubs and trees. A few are used in gardens; in particular, the marvel of Peru—*Mirabilis peruviana*, a colorful, tuberous perennial—and the woody climbing bougainvillea. Both are from South America.

## CALIFORNIA GENERA AND SPECIES

The region has eight native genera with 26 native species and three nonnative species.

*Abronia* (sand-verbena) are sprawling, succulent herbs that favor sand dunes. Their brightly colored, fragrant flowers are arranged in dense heads.

*A. maritima* (red sand-verbena) has maroon-red flowers and is common along beaches in southern California.

*A. umbellata* (pink sand-verbena; see fig. a) has pale pink or pink-purple flowers and occurs on coastal dunes along the entire coast.

*A. latifolia* (yellow sand-verbena) has bright yellow flowers and typifies northern beaches.

*A. villosa* (desert sand-verbena) colors sandy flats in the low deserts with rose-purple flowers.

*Boerhavia* (spiderlings) are sprawling to erect annuals and perennials with clusters of tiny white or red flowers. Most are desert-dwellers—some of which are summer annuals—but a few enter the dry mountains of southern California.

*Mirabilis* (four o'clocks; see fig. b) are bushy, sometimes woody perennials from taproots with leafy bracts surrounding one to several trumpet-shaped flowers that open in the evening. Three species are introduced; the others are native to drylands mostly in southern California and the deserts.

### a. *Abronia umbellata* (pink sand-verbena)

NYCTAGINACEAE (Four O'Clock Family)

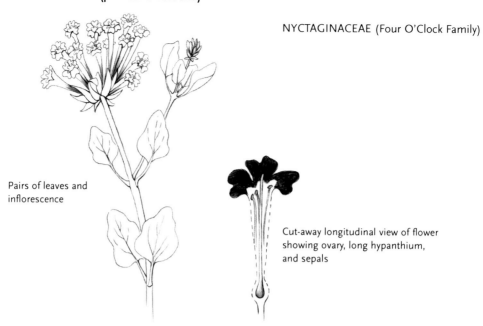

Pairs of leaves and inflorescence

Cut-away longitudinal view of flower showing ovary, long hypanthium, and sepals

### b. *Mirabilis multiflora* (desert four o'clock)

Leaves and flowers inside sepal-like bracts

# OLEACEAE (Olive Family)

RECOGNITION AT A GLANCE  Woody plants with (often) opposite leaves and flowers with 5 petals (when present) and only 2 stamens.

VEGETATIVE FEATURES  Shrubs and trees with (often) opposite, simple to pinnately compound leaves.

FLOWERS  Small, sometimes showy and bisexual; sometimes petalless, dioecious, and wind-pollinated.

FLOWER PARTS  2 or 5 tiny sepals, 2 or 5 petals fused to form a tube, 2 stamens joined to petal tube, and a single pistil with a superior, 2-chambered ovary and 2 stigma lobes.

FRUITS  Drupes, capsules, and samaras.

RELATED OR SIMILAR-LOOKING FAMILIES  The olive family is quite variable with regards to pollination (wind- or insect-pollinated) and leaf design (leaves are simple or compound). The best unifying feature is the stamen number.

STATISTICS  900 species in arid habitats worldwide; some also live in riparian areas. Besides the olive (*Olea europea*) with its wonderfully versatile fruits, the family is noted for such favorite ornamentals as jasmine (*Jasminum* spp.), lilacs (*Syringa* spp.), forsythia (*Forsythia* spp.), and ashes (*Fraxinus* spp.). Ash wood is the preferred material for baseball bats.

## CALIFORNIA GENERA AND SPECIES

The region has three native genera with eight species, and one nonnative genus and species.

*Fraxinus* (ashes) are trees or large shrubs with pairs of pinnately compound leaves and winged fruits.

*F. latifolia* (Oregon ash; see fig.) is a common riparian tree in northern California and has wind-pollinated, petalless flowers.

*F. velutina* (velvet ash) replaces Oregon ash in southern California and differs from it by having tiny stalks on the leaflets of each compound leaf.

*F. dipetala* (flowering ash) is a large shrub or small tree of the California foothills with white, two-petalled flowers.

*Menodora* (3 species) are herbaceous perennials or small shrubs mostly from deserts, with alternate leaves, dry capsules, and white or yellow flowers.

*Forestiera pubescens* (desert-olive) is a small tree from oases in hot, arid places, with clusters of leaves and tiny green, petalless flowers and purple drupes.

*Olea europea* (olive) is occasionally naturalized in California's foothills. It features tiny, whitish flowers and simple leaves.

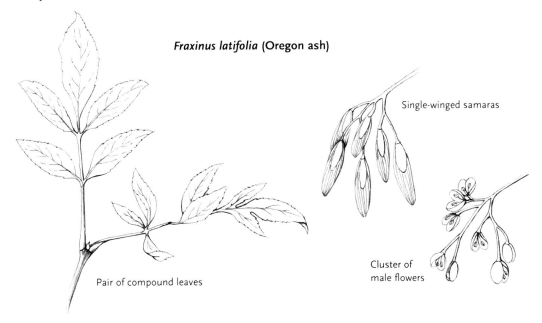

**Fraxinus latifolia (Oregon ash)**

Single-winged samaras

Pair of compound leaves

Cluster of male flowers

# ONAGRACEAE (Evening-Primrose Family)

RECOGNITION AT A GLANCE Herbaceous plants with simple leaves and (usually) showy flowers with 4 petals, 4 or 8 stamens, and an inferior ovary.

VEGETATIVE FEATURES Herbaceous annuals and perennials and simple, sometimes lobed, basal, alternate, or opposite leaves.

FLOWERS Usually showy, regular, cup- or saucer-shaped, and solitary or in racemes.

FLOWER PARTS (Usually) 4 separate sepals, (usually) 4 separate petals, and 4 or 8 stamens joined to a tubular hypanthium, and a single pistil with an inferior, usually 4-chambered ovary.

FRUITS (Usually) capsules with many seeds.

RELATED OR SIMILAR-LOOKING FAMILIES Only a few families consistently have flowers with parts in fours. The most diverse is the mustard family (Brassicaceae) and the closely related caper family (Capparidaceae). Those families have superior ovaries and usually 6 stamens.

STATISTICS 650 species widely distributed, with the greatest diversity in western North America and the mountains of tropical America. Many are garden ornamentals, the most popular of which is the genus *Fuchsia* from Mexico to South America. Other garden ornamentals include evening-primroses (*Oenothera* spp.), clarkias and godetias (*Clarkia* spp.), and California-fuchsia (*Epilobium canum*).

## CALIFORNIA GENERA AND SPECIES

The region has eight mostly native genera.

### UNUSUAL GENERA

*Ludwigia* (water-primroses) are marsh plants with creeping stems and 5-petaled, yellow flowers.

*Circaea alpina* (enchanter's nightshade) is a small perennial with broad leaves and tiny, 2-petaled flowers. It lives in moist, shaded places.

### MORE TYPICAL GENERA

*Gayophytum* are slender annuals from mountain forests with threadlike stems and tiny white flowers.

*Epilobium* (several different common names) are varied; some are weedy annuals, others are herbaceous or woody perennials with traveling roots; all feature racemes of pink, purple, red, or white flowers usually with hairy seeds.

*E. densifolia* is a widespread early summer annual with sticky leaves, pink-purple flowers, notched petals, and hairless seeds. It was formerly in the genus *Boisduvalia*.

*E. canum* (hummingbird- or California-fuchsia; see fig. a) is a woody perennial with narrow, hairy leaves, trumpet-shaped, scarlet flowers, and hairy seeds. It was formerly in the genus *Zauschneria*.

*E. angustifolium* (fireweed) is a rhizomatous perennial with narrow leaves, tall racemes of pink-purple flowers, unnotched petals, and hairy seeds.

*E. obcordatum* (rock-fringe) is a sprawling, high-mountain perennial with rounded, bluish green leaves; small clusters of large, magenta-pink flowers with notched petals; and hairy seeds.

*Camissonia* (suncups) are annuals and perennials with white, pink, or yellow petals; a ball-shaped stigma; and a long hypanthium. Most species open their flowers in the daylight hours. Many live in deserts. Camissonias were once considered part of the genus *Oenothera* but the stigmas differ.

*C. ovata* (golden eggs) is a rosetted coastal perennial with single yellow flowers and a very long hypanthium.

*C. cheiranthifolia* (beach suncups) is a sprawling coastal dune perennial with a series of bright yellow flowers on radiating stems.

*C. boothii* (bottle scrubber) is an annual with white to pink flowers and woody seed pods that resemble a coarse bottle scrubber.

*Oenothera* (evening-primroses) are annuals and perennials with large yellow, white, or pinkish flowers that open in the evening and early morning and have a 4-lobed, crosslike stigma. Most are pollinated by hawk-moths and hummingbirds.

*O. elata* var. *hookeri* (Hooker's evening-primrose) is a stout biennial with tall racemes of large yellow flowers. It is widespread in moist gullies.

*O. deltoides* (Antioch Dunes evening-primrose; see fig. b) is a rosetted annual with large, fragrant white flowers that fade pink. It mostly lives in sandy places in the deserts.

*Clarkia* (clarkia, godetias) are annuals with sometimes clawed, purple, pink, or white petals and sepals that are turned to one side or partially stuck together. Several species are rare. The clarkias fall into two groups, described below.

### SPECIES WITH BOWL- OR CUP-SHAPED FLOWERS (FORMERLY GODETIA)

To identify species, note such features as number of stamens, number of ridges on the ovary, and whether buds are drooping or erect.

*Clarkia gracilis* (graceful clarkia) and *C. amoena* (farewell-to-spring) have nodding flower buds and are widely distributed.

*C. purpurea* (winecup clarkia) is highly variable but usually has small flowers, stiff flower buds, and petals from purple to deep wine-red.

### SPECIES WITH FAN-SHAPED PETALS AND FLAT FLOWERS

*Clarkia unguiculata* (elegant clarkia; see fig. c) has tall stems with pink, unlobed petals. It is widespread in woodlands.

*C. biloba* has oversized pink, deeply notched petals, and occurs in inland foothills.

*C. rhomboidea* (diamond clarkia) has diamond-shaped, pink-purple, often spotted petals and lives in open forests in the north.

*C. concinna* (red ribbons clarkia) is a low annual with brilliant pink petals lobed into 3 segments; it lives in woodlands and forests in the foothills.

*C. breweri* (fairy fans) is a small annual with fragrant, pale pink petals split into 3 lobes, and lives on rock scree in the inner central Coast Ranges.

**a. *Epilobium canum* (California-fuchsia)**

ONAGRACEAE (Evening-Primrose Family)

**b. *Oenothera deltoides* var. *howellii* (Antioch Dunes evening-primrose)**

Cut-away view of flower showing 4 sepals and petals, hypanthium, 8 stamens, and a single style

Wavy leaves and flowers

**c. *Clarkia unguiculata* (elegant clarkia)**

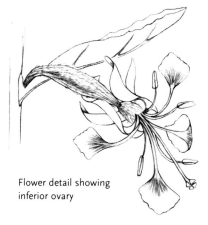

Flower detail showing inferior ovary

## ORCHIDACEAE (Orchid Family)

RECOGNITION AT A GLANCE Perennials with simple, parallel-veined leaves and irregular flowers with an enlarged lower lip and stamen(s) fused to the style and stigma as a column.

VEGETATIVE FEATURES Herbaceous perennials from rhizomes, tubers, and corms or fleshy roots connected to special soil fungi; and simple, entire, parallel-veined leaves (reduced to scales in the parasitic species).

FLOWERS Range from small to large and showy, are irregular with a definite lower lip, and are solitary or arranged in spikes and racemes.

FLOWER PARTS 3 similar, often petal-like sepals; 3 petals, the lower enlarged into an intricate lip with a landing platform and nectar guides; 1 (rarely 2) stamen(s) fused to the style and stigma to form a column; and a single, inferior ovary.

FRUITS Capsules with thousands of tiny, dustlike seeds.

RELATED OR SIMILAR-LOOKING FAMILIES The orchid family is a typical monocot family in terms of flower parts and leaf-vein pattern, but differs from the lilies (Liliaceae) and irises (Iridaceae) by having irregular flowers with the lower petal enlarged into a lip. This design is unlikely to be confused with any other family. The orchids represent a highly specialized evolutionary line, where the great diversity of flower design results in a huge array of different, often specialized pollinators. Other specializations include seeds reduced to a minimum size without stored food. The seeds depend on special mycorrhizal soil fungi to germinate.

STATISTICS The world's largest flowering plant family with somewhere between 20,000 and 30,000 species. The family is widely distributed and found on every continent except Antarctica, but the greatest diversity by far is in the tropics. Many tropical species live as epiphytes perched in trees. The plants are of three basic types: epiphytes, terrestrial, and fungus parasites without green leaves. Besides the genus *Vanilla*, which provides us with the flavoring of that name, there are hundreds of cultivated species plus a tremendous number of hybrids and selected cultivars. Prominent ornamental genera include *Cattleya, Cymbidium, Oncidium, Vanda, Dendrobium, Phalaenopsis, Paphiopedilum, Epidendrum*, and many more.

### CALIFORNIA GENERA AND SPECIES

The region has 11 native genera and 31 species.

*GENERA WITH MANY SMALL WHITE OR GREENISH FLOWERS IN DENSE SPIKES OR RACEMES*
*Platanthera* (rein orchids) live in wet meadows, and have green leaves at flowering time, and green or white flowers with a tapered, nectar-containing spur.

*P. leucostachys* (snowy rein orchid) has snowy white flowers.

*P. sparsiflora* (green rein orchid) has green flowers.

*Piperia* (also rein orchids) live in dry woods and coastal meadows, have leaves that turn brown by flowering time, and have green or white flowers with a spur. The several species are difficult to sort out.

*Spiranthes* (ladies' tresses) have white to cream-colored flowers without a spur, arranged in a spiralled spike. They live in wet meadows.

*Goodyera oblongifolia* (rattlesnake orchid) has broad, bluish green leaves netted with white veins, and white flowers arranged in a spiral spike. It lives in conifer woods.

*GENERA WITHOUT GREEN LEAVES*
Leaves are reduced to brownish or reddish scales.

*Corallorhiza* (coralroots) have branched, coral-like roots and racemes of red or purple flowers with a lower lip often striped or spotted.

*C. maculata* (spotted coralroot; see fig. a) is widespread in conifer forests and has a spotted lip.

*C. striata* (striped coralroot) lives in conifer forests and has a lip with red-purple stripes.

*Cephalanthera austinae* (ghost or phantom orchid) lives in conifer forests and has pure white stems and flowers with a patch of yellow on the lip.

GENERA WITH LARGER, USUALLY SHOWY FLOWERS
*Calypso bulbosa* (fairy slipper) grows in leaf mold and decomposing wood in conifer forests. It has a single broad leaf and a single exquisite, fragrant, pink-purple flower resembling a miniature corsage orchid.

*Cypripedium* (lady slippers) live in conifer woods or bogs. They feature a few broad leaves and racemes of curious flowers with the lip enlarged into an inflated slipper.

*C. californicum* (California ladyslipper; see fig. b) lives in northern bogs and has a white lip and pale yellow lateral petals.

*C. montanum* (mountain ladyslipper) lives in dry conifer forests and has few large flowers with a fragrant white lip and twisted brown lateral petals.

*C. fasciculatum* (brownies) lives in conifer forests and has a pair of broad leaves and a cluster of somewhat smaller flowers with a greenish lip and reddish brown lateral petals.

*Epipactis gigantea* (brook orchid; chatterbox) lives along wooded streams and has racemes of flowers with a movable lip and subtle colors including pink, yellow, green, and brown.

*E. helleborine* (helleborine) is an introduced European orchid widespread in the Bay Area with smaller, red-purple flowers.

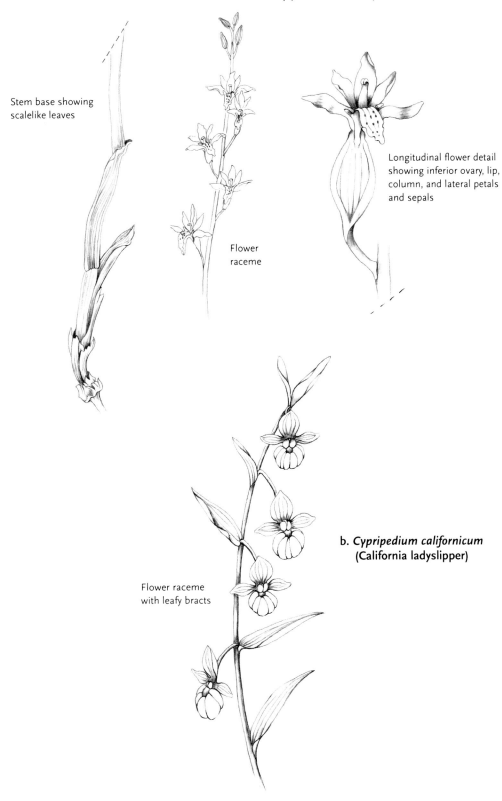

# PAPAVERACEAE (Poppy Family)

**RECOGNITION AT A GLANCE** Flowers with sepals that pop off as petals open (most), and sepals that are half the number of petals.

**VEGETATIVE FEATURES** Herbaceous annuals, perennials, and small shrubs with caustic clear or orange sap often containing opiates. Variable leaves; leaves in several genera are highly dissected and fernlike.

**FLOWERS** Usually showy, shallowly saucer-shaped or irregular and like bleeding hearts, and often solitary.

**FLOWER PARTS** 2 or 3 separate sepals that (often) fall away as flowers open, 4 or 6 separate regular petals or 4 irregular, partly fused petals in 2 pairs (each pair is different from the other), 6 to numerous stamens, and a single pistil with a superior ovary. Those flowers with numerous stamens lack nectar and offer pollen as a reward to bees; those with 6 stamens and irregular flowers produce nectar, and the larger pair of petals is saclike or spurred at the base.

**FRUITS** Capsules, sometimes discharging seeds explosively.

**RELATED OR SIMILAR-LOOKING FAMILIES** Other families with numerous stamens and wide open flowers often have several separate pistils or, for those with a single pistil, sepals that remain on the flower after it has opened. The fumitory family (Fumariaceae) is included in the Papaveraceae because of the similar chemistry of its poisonous compounds and similar leaf design. That branch of the poppy family is distinguished by its irregular, often spurred flowers, fixed number of stamens, and petals that are partly joined.

**STATISTICS** 300 species mostly in the northern hemisphere; a few live in the highlands of the tropics. The opium poppy (*Papaver somniferum*) has a long history of use for its opium-derived compounds and its seeds, which are widely used in baking. The family is also noted for many handsome garden flowers including poppies (*Papaver* spp.), California poppy (*Eschscholzia californica*), blue poppy (*Meconopsis* spp.), bleeding hearts (*Dicentra* spp.), and *Corydalis*.

## CALIFORNIA GENERA AND SPECIES

The region has 12 native genera and two non-native genera.

*GENERA WITH IRREGULAR FLOWERS, PERSISTENT SEPALS, AND 6 STAMENS*

*Dicentra* (bleeding heart, golden eardrops, and others) are rhizomatous perennials with highly divided, fernlike leaves and irregular, heart-shaped pink, peach-colored, creamy, or yellow flowers.

*D. formosa* (western bleeding heart) is a common, pink-flowered ground cover in conifer forests.

*D. chrysantha* (golden eardrops) is a tall, sun-loving plant with yellow flowers that appears in quantity after fire in chaparral.

*Corydalis* are annuals and perennials with highly divided, fernlike leaves and irregular flowers with a nectar spur.

*C. caseana* has pale pink flowers and reaches 4 feet in height; it is found in wet meadows and brushy streams in the Cascade Mountains.

*Fumaria* (fumitories) are introduced annuals with feathery foliage and narrow, irregular flowers. They have naturalized in coastal forests near the Bay Area.

*GENERA WITH REGULAR FLOWERS, SEPALS THAT FALL AWAY, AND (USUALLY) NUMEROUS STAMENS*

**GENERA THAT ARE BUSHY OR WOODY** *Romneya* (matilija poppy; see fig. a) are woody-based perennials with tall stems and very large flowers (perhaps the largest in California) with crumpled, white petals. They live in burned-over areas in southern California's chaparral.

*Dendromecon* (bush poppies) are small shrubs with simple, oblique leaves and bright yellow flowers. They live on the edge of chaparral and sprout from seed by the hundreds after brush fires.

*Argemone* (prickly poppies) are perennials with spiny stems, leaves, and sepals, and large white flowers. They live in arid places and deserts.

GENERA THAT ARE ENTIRELY HERBACEOUS

*Eschscholzia* (California poppies) are (mostly) annuals with finely divided, fernlike leaves and yellow to bright orange flowers.

*E. californica* (see fig. b), our state flower, is widely distributed through California's foothills. Its flowers have a red rim under the petals.

*Platystemon californicus* (cream cups) is an annual with small, cream-colored flowers often blotched yellow and with flattened stamens. It is abundant in the foothills.

*Meconella* are diminutive annuals with tiny, pure white flowers. They are widely scattered in the foothills.

*Stylomecon heterophylla* (wind poppy) is a woodland annual with variable leaves and wide open, dark orange flowers with a purple center. It lives in foothill woodlands.

*Papaver californicum* (flame or fire poppy) is a woodland annual that appears after fire. It features large orange flowers similar to wind poppy but with a yellow-green center and no style on the ovary.

**a. *Romneya coulteri* (matilija poppy)**

Flower detail showing crumpled petals and numerous stamens

**b. *Eschscholzia californica* (California poppy)**

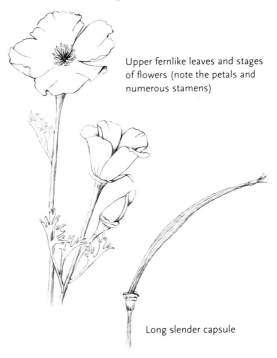

Upper fernlike leaves and stages of flowers (note the petals and numerous stamens)

Long slender capsule

PAPAVERACEAE (POPPY FAMILY) 123

# PINACEAE (Pine Family)

**RECOGNITION AT A GLANCE** (Mostly) evergreen trees with needles shed singly or attached to tiny spur shoots, and seed cones with papery or woody spirally arranged scales.

**VEGETATIVE FEATURES** Monoecious, evergreen shrubs or trees (some nonnatives are deciduous) with spirally arranged needles that fall singly or are attached to tiny spur shoots.

**FLOWERS** Missing; pines reproduce by pollen and seed cones.

Seed cones: small to substantially large seed cones with papery or woody, spirally arranged scales and bracts (bracts are often hidden); each scale bears 2 seeds.

**RELATED OR SIMILAR-LOOKING FAMILIES** The pine family, with its needled branches and the spirally arranged scales of the seed cones, bears a resemblance to the redwood family (Taxodiaceae), but that family sheds its needles attached to twigs. The yew family (Taxaceae) also has needles but the plants are dioecious and the seed cones consist of a single seed enclosed in a fleshy aril.

**STATISTICS** This is the world's largest conifer family, with around 200 species across the northern hemisphere, and is diverse in the mountains of Mexico, California, China, and the Himalayas. Many dominate mountain and subarctic forests over vast tracts of land. Several are valuable timber trees, including the white pines and Douglas-fir. Many are also used in parks and gardens as handsome specimen trees; most of the 10 genera are widely cultivated.

## CALIFORNIA GENERA AND SPECIES

The region has five native genera.

*Pinus* (pines) has needles on tiny spur shoots in a fixed number: 1, 2, 3, 4, or 5. The seed cones usually have woody scales and vary greatly in size and other characteristics. Around 19 species in California that belong to several different subgroups.

*P. ponderosa* (ponderosa pine; see fig. a) is a substantial tree at middle elevations in the mountains, whose textured bark resembles jigsaw puzzle pieces. It has 3 needles per spur shoot, and seed cone scales that end in a small, outturned prickle.

*P. jeffreyi* (Jeffrey pine) is similar to ponderosa pine but occurs at higher elevations or on drier sites, has fragrant bark, and seed cones to 8 inches long with scales that have inturned prickles.

*P. lambertiana* (sugar pine) is a very tall mountain tree often associated with ponderosa pine and identified by pendulous seed cones to 20 inches long and papery scales without prickles.

*P. sabiniana* (gray or foothill pine) is an often multi-trunked tree from the hot foothills, with wispy, gray needles in threes and heavy seed cones with stout scales that end in a substantial spine.

*P. contorta* subsp. *murrayana* (lodgepole pine) is a tree from the subalpine zone with scaly bark, short needles in twos, and small seed cones with a prickle on each scale.

*P. muricata* (bishop pine) and *P. radiata* (Monterey pine) are a pair of pines from coastal bluffs with permanently attached seed cones that open after fire. Bishop pine has needles in twos; Monterey has needles in threes.

*P. longaeva* (bristlecone pine) is a very slow-growing, extremely long-lived tree (up to nearly 5,000 years) with tight clusters of 5 needles and seed cones with bristle-tipped scales. The best development of this picturesque tree is in the White Mountains on dolomite soils near the California-Nevada border.

*Abies* (true firs) have needles on branches that leave a smooth scar when they drop, and upright, candlelike seed cones that shatter when they are ripe.

*A. concolor* (white fir; see fig. b) lives in the ponderosa pine belt, has whitish bark and relatively long needles that lie flat on the lower branches.

*A. magnifica* (red fir) lives in the high mountains, has red-brown bark, and short, stiff, upright needles.

*A. grandis* (grand fir) lives on the north coast and has blunt, glossy green needles of two lengths.

*A. bracteata* (Santa Lucia fir) is restricted to limestone scree in the Santa Lucia Mountains. It features long, spine-tipped, glossy needles, and seed cones with protruding bracts between the scales.

*Picea* (spruces) have needles that leave behind a peg when they drop, and hanging seed cones with papery, often fluted or toothed scales. The cones fall intact.

*P. sitchensis* (Sitka spruce) is a substantial tree in north coastal forests with scaly bark, prickly dark green needles, and fluted seed cone scales.

*Tsuga* (hemlocks) have needles on branches that leave a rough place when they drop, and hanging seed cones with papery scales. Their leader is gracefully drooping, not stiff like spruces and firs.

*T. heterophylla* (coast hemlock) lives in north coastal forests and has short, light green needles of two lengths, and tiny seed cones.

*T. mertensiana* (mountain hemlock) lives near timberline in the high mountains and has short, spirally arranged, bluish green needles and relatively long seed cones.

*Pseudotsuga* (Douglas-firs) have needles that leave a rough place when they drop and have hanging seed cones with more-substantial scales and protruding 3-pronged bracts.

*P. menziesii* (Douglas-fir; see fig. c) lives in a variety of conifer and mixed forests, has rough, craggy bark, needles all around the twigs, and seed cones to 3 inches long. It is highly variable and a major source of lumber.

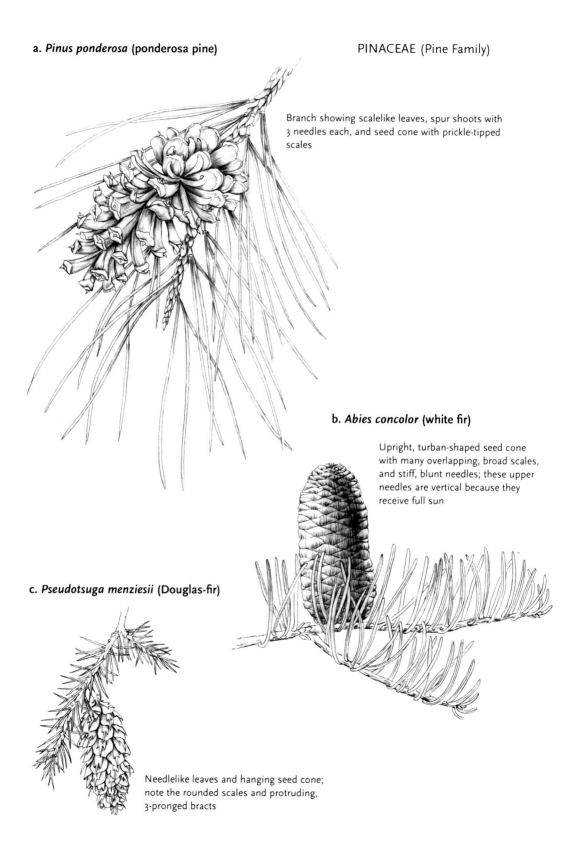

**a. *Pinus ponderosa* (ponderosa pine)**  PINACEAE (Pine Family)

Branch showing scalelike leaves, spur shoots with 3 needles each, and seed cone with prickle-tipped scales

**b. *Abies concolor* (white fir)**

Upright, turban-shaped seed cone with many overlapping, broad scales, and stiff, blunt needles; these upper needles are vertical because they receive full sun

**c. *Pseudotsuga menziesii* (Douglas-fir)**

Needlelike leaves and hanging seed cone; note the rounded scales and protruding, 3-pronged bracts

# PLATANACEAE (Plane Tree Family)

**RECOGNITION AT A GLANCE** Deciduous trees with maplelike leaves, collarlike stipules, mottled bark, and ball-shaped clusters of dry fruits.

**VEGETATIVE FEATURES** Deciduous, monoecious trees with puzzlelike bark and alternate, palmately lobed, maplelike leaves, sometimes with conspicuous, collarlike stipules. Leaf stalks are hollow at the base and fit over the lateral buds.

**FLOWERS** Small, unisexual, greenish, wind-pollinated, and in tightly packed, globe-shaped heads arranged in hanging racemes.

**FLOWER PARTS** Male flowers have a few sepals and stamens; female flowers a few sepals and a single pistil with a superior ovary.

**FRUITS** Fuzzy, 1-seeded achenes in dense, globe-shaped clusters that break apart when ripe.

**RELATED OR SIMILAR-LOOKING FAMILIES** The plane tree family is closely related to the non-native witch-hazel family (Hamamelidaceae) but differs in its leaf design and flower arrangement. Because of the large, maplelike leaves, the family might be mistaken for the maples (Aceraceae), but that family has leaves in pairs, 2-winged samaras, and lacks conspicuous stipules.

**STATISTICS** Around eight species in the northern hemisphere, often growing near watercourses. A few are cultivated as street and park trees, including the popular London plane tree (*Platanus* x *acerifolia*).

## CALIFORNIA GENERA AND SPECIES

The region has one native species.

*Platanus racemosa* (western sycamore; see fig. a) is a large tree found in riparian forests from the northern Sierra foothills and central Coast Ranges south. They reach their best development in southern California coastal canyons, where they join company with California bay (*Umbellularia californica*), white alder (*Alnus rhombifolia*), coast live oak (*Quercus agrifolia*), and southern black walnut (*Juglans californica* var. *californica*).

**a. *Platanus racemosa* (western sycamore)**

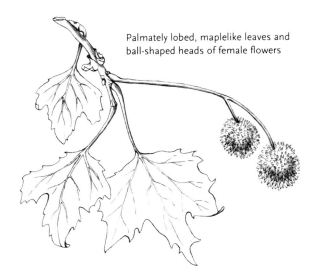

Palmately lobed, maplelike leaves and ball-shaped heads of female flowers

Female flower detail showing tuft of hairs and single pistil

# POACEAE (Grass Family)

**RECOGNITION AT A GLANCE** Grasslike plants with round, hollow stems; 2 rows of leaves; tiny, bisexual flowers without petals arranged in spikelets; and fruits called grains.

**VEGETATIVE FEATURES** Herbaceous annuals or perennials with densely fibrous roots and often underground stolons or rhizomes. The round, hollow stems bear 2 rows of leaves (or leaves are basal). Leaves consist of a rolled or flat blade with a base that forms a sheath around the stem. Leaves usually have pairs of tiny, earlobe-shaped appendages called auricles.

**FLOWERS** Tiny, greenish, (usually) bisexual, wind-pollinated, petalless, and arranged in tiny spikelets with (usually) 2 bracts (glumes) at the base. The spikelets themselves are arranged in heads, panicles, spikes, or racemes. Spikelets may have several flowers (florets), a single floret, or some fertile florets and some sterile florets (no stamens or pistil).

**FLOWER PARTS** Each floret consists of a pair of bracts, the lemma (away from the flowering stalk or rachis) and the palea (next to the rachis); 3 stamens; and a single pistil with a superior ovary and 2 feathery stigmas. Glumes, lemmas, and paleas often have spikelike extensions called awns.

**FRUITS** 1-seeded caryopses (grains).

**RELATED OR SIMILAR-LOOKING FAMILIES** Grasses look superficially like several other monocot families, the most prominent of which are the sedges (Cyperaceae) and rushes (Juncaceae). They can be told apart both on the basis of leaves, stems, flowers, and fruits. Sedges usually have triangular, solid stems and 3 rows of channeled leaves; rushes have round, pithy stems and varied leaves (but no auricles). Sedge flowers are often unisexual and often have bristles; rush flowers are bisexual but have a definite perianth of 6 tepals.

**STATISTICS** With 10,000 or more species worldwide, the grasses are the second largest monocot family after the orchids (Orchidaceae). Besides dominating vast areas of grassland, prairie, and meadow, the grasses have numerous important economic uses not surpassed by any other family. Woody grasses known as bamboos are popular ornamentals and have numerous uses in eastern Asia. Many other genera are used in gardens as ornamentals or to create the turf we call lawns. All of the major grain crops belong to this family, including corn, barley, wheat, millet, rice, sorghum, and oats. Sugar is extracted from the tropical sugar cane as well as from corn. And grasses provide the basic food for grazing animals such as cows and sheep.

## CALIFORNIA GENERA AND SPECIES

The grass family is our second most prominent family, with hundreds of species. Many are native, but a considerable number are introduced from Europe, Asia, and Africa. The basic forms include annuals, clump-forming perennial bunchgrasses, rhizomatous grasses with widely spreading rhizomes, and turflike grasses with dense, short rhizomes. Because of its size, it is most convenient to divide the family into tribes. A sampling follows.

### AGROSTIDEAE (BENT-GRASS TRIBE)

One-flowered spikelets in racemes or panicles. Some genera include:

*Polypogon monspeliensis* (rabbitfoot grass), nonnative annuals with dense, furry, spikelike panicles.

*Nassella* (see fig. a), *Hesperostipa*, and *Achnatherum* (formerly all in the genus *Stipa*, needlegrasses) are bunchgrasses with panicles of florets bearing a bent or twisted awn.

*Phleum* (timothies) are robust perennials with dense, cylinder-shaped, compound spikes.

*Calamagrostis* (reedgrasses) are canelike or clumped bunchgrasses with panicles.

*Muhlenbergia* (muhlies) are varied rhizomatous grasses and bunchgrasses.

*Sporobolus* (alkali sacaton) is a coarse bunchgrass with open panicles and divergent branchlets.

*Ammophila arenaria* (dune grass) is a nonnative, densely rhizomatous grass that has crowded out our native dune plants.

PHALARIDEAE (CANARY GRASS TRIBE)
Similar to Agrostideae *except* there are sterile flower parts below the single functional flower of each spikelet.

*Hierchloe occidentalis* (vanilla grass; see fig. b) is a common shade-loving grass of moist northern forests, with sweetly fragrant leaves.

AVENEAE (OAT TRIBE)
Two- to several-flowered spikelets in racemes or panicles; spikelets that fall apart above the glumes; and a lemma with a bent awn attached to its back (not its tip).

*Holcus lanatus* (velvet grass) is a European perennial with velvety leaves and dense panicles of pinkish florets.

*Danthonia* (oat-grasses) are small native bunchgrasses with horizontally or obliquely trending flowering culms.

*Koeleria macrantha* (June grass) is a widespread, small bunchgrass with interrupted spikes.

*Deschampsia* (tufted hairgrasses) are densely tufted, native bunchgrasses.

*Aira caryophylla* (hairgrass) is a small, nonnative annual with threadlike stalks and tiny spikelets.

*Avena* (wild oats) are invasive, European annuals with panicles of nodding spikelets. Accidentally introduced with the cultivated oat, wild oats now dominate huge areas of grasslands throughout our foothills.

FESTUCEAE (FESCUE TRIBE)
The features are similar to the oat tribe except the awn is straight and attached to the end of the lemma.

GENERA WITH A BAMBOOLIKE APPEARANCE
*Cortaderia* (pampas grasses) are huge, tufted South American perennials with woody rhizomes and immense white or pinkish, plumelike panicles. They have become a major problem in coastal plant communities.

*Arundo donax* (giant reed) is a 12- to 15-foot bamboolike African grass that has invaded many wetlands.

*Phragmites australis* (native reed) is another bamboolike plant, found along waterways in arid mountains.

GENERA OF SMALLER STATURE *Briza* (rattlesnake grasses) are Mediterranean annuals with spikes that resemble rattlesnake tails.

*Dactylis glomerata* (orchard grass) is a small European perennial naturalized in the mountains.

*Bromus* (bromes) include native perennials and mostly nonnative, highly invasive annuals with dense spikes with sharp, raspy awns.

*Festuca* (fescues; see fig. c) include natives and nonnatives that are usually bunchgrasses with less-spiny awns.

*Melica* (melics) are mostly small, native bunchgrasses with short spikelets arranged in spikes or panicles, sometimes with awns.

*Poa* (bluegrasses) are native and nonnative annuals and perennials with awnless florets.

CHLORIDEAE (GRAMMA GRASS TRIBE)
Small spikelets are arranged along one side of the main stalk and directly attached to it. The spikes are usually branched. A few genera:

*Spartina foliosa* (cord grass) is a tall, canelike grass lining tidal channels in salt marshes.

*Cynodon dactylon* (Bermuda grass) is a tough, invasive, nonnative, rhizomatous weed with wiry, leafy stems.

*Distichlis spicata* (salt grass; see fig. d) is a similar-looking sprawling native perennial with creeping stems and 2-ranked leaves typical of salt marshes.

*Bouteloua* (gramma grasses) are small, tufted, turf-forming grasses from high deserts.

HORDEAE (BARLEY OR FOXTAIL TRIBE)
Spikelets are attached directly to both sides of a single, usually large spike, and the main stalk of the spike ascends in zigzag fashion.

*Hordeum* (foxtail, wild barley) are weedy nonnative and native annuals and perennials with sharp awns that lodge in clothing and fur.

*Lolium* (European rye-grasses) are aggressive annual and perennial weeds often with flattened spikes.

*Leymus* and *Elymus* (native rye-grasses and others; see fig. e) are similar-looking, mostly clumped to rhizomatous perennials with (usually) leafy culms and flattened spikes of spikelets.

PANICEAE (PANIC GRASS TRIBE)
The spikelets and glumes both fall, and the lemmas and paleas are thick, often leathery, and have a different appearance from the glumes.

*Digitaria* (crab grass) is an infamous, non-native, summer weed with narrowly lance-shaped leaves.

*Pennisetum* (fountaingrasses) are African perennials, often used in gardens, that have escaped and become a menace in southern California. They are medium-sized bunchgrasses with long, sometimes red bristles mixed with the spikelets.

*Panicum* (panic grasses; see fig. f) are native perennials found on seeps, with divergent stems that carry lance-shaped leaves and panicles of tiny spikelets.

### AGROSTIDEAE (BENT-GRASS TRIBE)

#### a. *Nassella pulchra* (purple needlegrass)

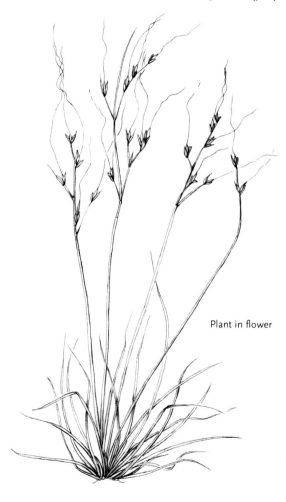

Plant in flower

Detail of fruiting spikelets showing glumes and caryopsis with twisted awns

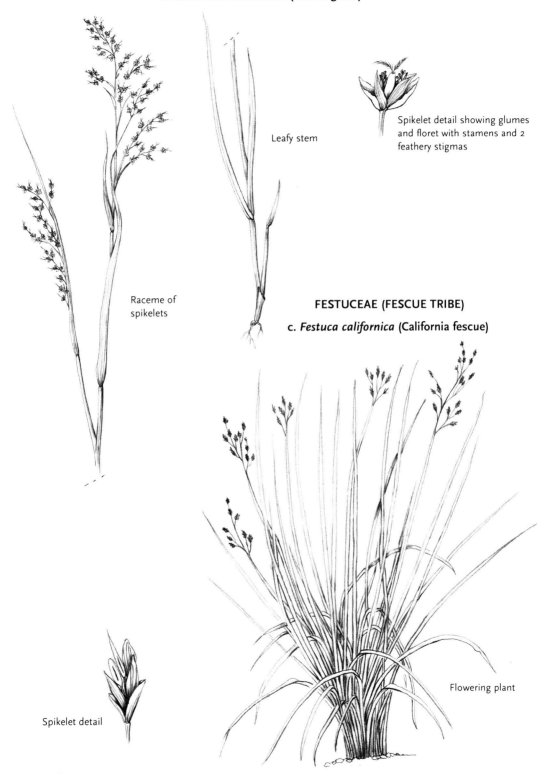

## CHLORIDEAE (GRAMMA GRASS TRIBE)
### d. *Distichlis spicata* (salt grass)

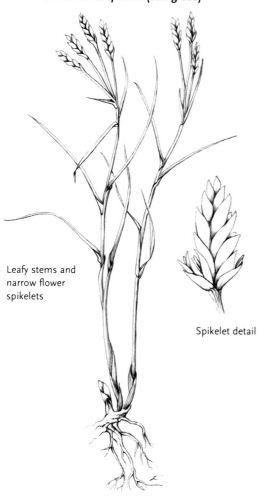

Leafy stems and narrow flower spikelets

Spikelet detail

## HORDEAE (BARLEY OR FOXTAIL TRIBE)
### e. *Elymus elymoides* (squirrel tail grass)

Leaves and tail-like spike

Spikelet detail showing long awns on glumes and florets

## PANICEAE (PANIC GRASS TRIBE)

### f. *Panicum thermale* (hot springs panic grass)

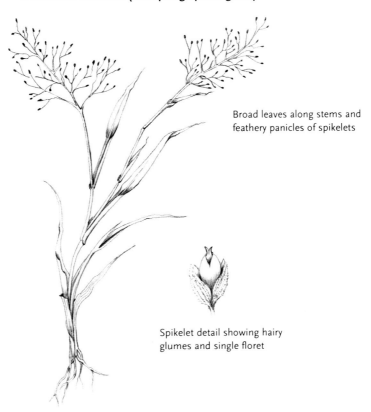

Broad leaves along stems and feathery panicles of spikelets

Spikelet detail showing hairy glumes and single floret

# POLEMONIACEAE (Phlox Family)

**RECOGNITION AT A GLANCE** Herbaceous plants and small shrubs with tubular, pinwheel-like petals joined to a tube, and 3 stigmas.

**VEGETATIVE FEATURES** Herbaceous annuals and perennials, woody cushion plants, and small shrubs with alternate or opposite, simple, palmately divided, or pinnately compound leaves.

**FLOWERS** Small to medium-sized, showy, tubular with pinwheel-like petal lobes; solitary or arranged in dense cymes, panicles, and heads.

**FLOWER PARTS** 5 partly fused sepals often edged by translucent membranes, 5 petals fused to form a tube, 5 stamens attached to the tube, and a single pistil with a superior, 3-chambered ovary and 3 stigma lobes.

**FRUITS** Capsules with many seeds.

**RELATED OR SIMILAR-LOOKING FAMILIES** The overall number of flower parts and design in the phlox family may suggest other similar and sometimes related dicot families, including the waterleaf family (Hydrophyllaceae) and borage family (Boraginaceae). Those two families usually have their flowers in coiled cymes that unwind as flower buds open; waterleafs have 2-chambered ovaries and 2 stigmas; borages, 4-lobed ovaries containing 1 seed per lobe, and a single style.

**STATISTICS** 320 species, with a wide distribution in the northern hemisphere and great diversity in western North America and South America. Among the several garden ornamentals are cup-and-saucer vine (*Cobaea scandens*), phloxes (*Phlox* spp.), magic plant (*Cantua* spp.), and gilias (*Gilia* spp.).

## CALIFORNIA GENERA AND SPECIES

The region has 13 native genera.

### GENERA WITH PINNATELY COMPOUND LEAVES

*Polemonium* (Jacob's ladder; sky pilot; see fig. a) are herbaceous perennials with compound leaves similar to pea leaves, and clusters of blue or pinkish flowers. Most live in the mountains.

*Gilia* (gilias) are annuals in foothills and deserts, with alternate, pinnately compound or simple leaves and varied arrangements of blue, purple, white, or pink flowers. The many species are difficult to key out.

*G. capitata* (globe gilia) has pinnately compound leaves and globe-shaped heads of blue flowers.

*G. tricolor* (birdseye gilia; see fig. b) is a low-growing plant with pinnately divided leaves and small clusters of flowers with pale purple petal lobes, a dark purple base, and a yellow throat.

*G. leptalea* is a small mountain annual with simple leaves and magenta flowers.

*Ipomopsis* (scarlet-gilia; woolly heads) are biennials or perennials with thickened teeth along the leaf margins and usually very long, tubular flowers.

*Navarretia* spp. are annuals in vernal pools and hard-packed soils, with deeply dissected, sometimes skunk-scented leaves and spiny heads of white, yellow, purple, or blue flowers.

### GENERA WITH SIMPLE LEAVES

*Collomia* spp. are annuals with tight heads of flowers surrounded by bracts. Most species have inconspicuous purplish or pink flowers.

*C. grandiflora* (bride's bouquet) has large, apricot-colored flowers and blue pollen.

*Allophyllum* spp. are small annuals with usually palmately lobed upper leaves and small blue, purple, white, or red-purple flowers.

*Phlox* (phloxes) are usually woody-based perennials often forming mounds or mats with simple, sometimes needlelike leaves and showy pink, purple, or white flowers. The stamens are of uneven lengths.

*P. diffusa* (mat phlox) makes cushions with profuse, perfumed flowers in several colors. It lives in rocky places in the high mountains.

*P. speciosa* (showy phlox) is a sprawling perennial with lance-ovate leaves and clusters of large pink flowers with notched petals.

*P. gracilis* (annual phlox) is a small woodland annual with minute, pinkish flowers.

### GENERA WITH SPINY SEPALS AND SOMETIMES SPINY BRACTS

*Eriastrum* spp. (woolly stars) are dryland annuals and perennials with often spiny leaves, spiny bracted heads with dense woolly white hairs, and tubular blue, yellow, or purple flowers.

*Langloisia* (sunbonnets) and *Loeseliastrum* (calico flowers) are desert annuals.

### GENERA WITH DIGITATELY DIVIDED OR COMPOUND LEAVES

*Linanthus* spp. (see fig. c) are annuals in foothills, mountains, and deserts, with pairs of digitately compound leaves and heads of tubular yellow, white, pink, lavender, or purple flowers.

*Leptodactylon* (prickly-phloxes) are small shrubs or woody matted perennials with usually alternate, digitately divided, spine-tipped leaves and white, lavender, or pink flowers.

*L. californicum* is a small shrub from the mountains of southern California with spectacular pink flowers.

*L. pungens* is a dwarf, matted, high-mountain shrub with white to pale lavender flowers.

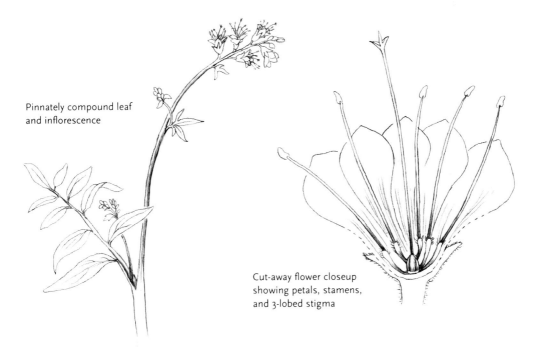

a. *Polemonium occidentale* (Jacob's ladder)

Pinnately compound leaf and inflorescence

Cut-away flower closeup showing petals, stamens, and 3-lobed stigma

## POLEMONIACEAE (Phlox Family)

### b. *Gilia tricolor* (birdseye gilia)

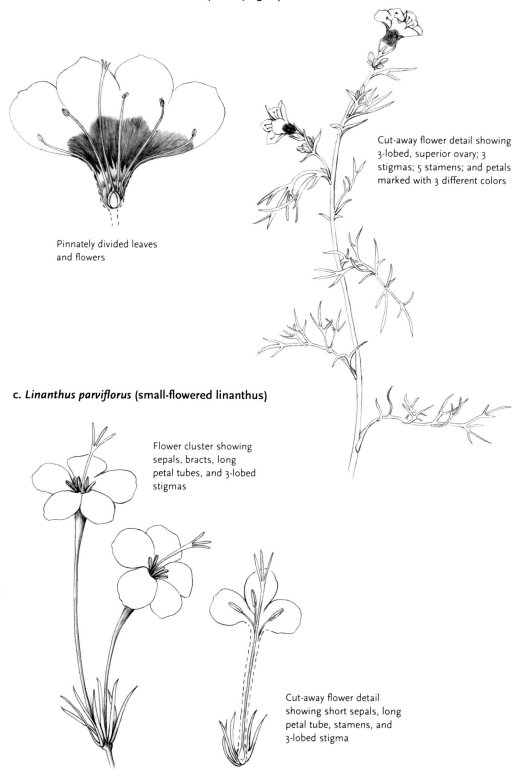

Pinnately divided leaves and flowers

Cut-away flower detail showing 3-lobed, superior ovary; 3 stigmas; 5 stamens; and petals marked with 3 different colors

### c. *Linanthus parviflorus* (small-flowered linanthus)

Flower cluster showing sepals, bracts, long petal tubes, and 3-lobed stigmas

Cut-away flower detail showing short sepals, long petal tube, stamens, and 3-lobed stigma

# POLYGONACEAE (Buckwheat Family)

RECOGNITION AT A GLANCE Mostly herbaceous plants with simple leaves attached to swollen nodes (or basal), sometimes papery stipules, and tiny flowers with 4 to 6 tepals arranged in dense clusters.

VEGETATIVE FEATURES Herbaceous annuals, perennials, and small shrubs with jointed stems; simple, untoothed, alternate or basal leaves attached to swollen nodes, and sometimes papery stipules (ochrea).

FLOWERS Tiny, green, brownish, white, yellow, pink, or red, and clustered in heads, spikes, or racemes sometimes inside cup- or vase-shaped bracts (involucres).

FLOWER PARTS 4 to 6 tepals (no distinction between sepals and petals), 3 to 9 stamens, and a single pistil with a superior, 3-sided ovary and 3 styles.

FRUITS Hard, usually 3-sided achenes.

RELATED OR SIMILAR-LOOKING FAMILIES The floral plan of a typical flower is different from most other families, especially those species that have 6 tepals, 9 stamens, and a 3-sided ovary. Nonetheless, some species that feature green, wind-pollinated flowers might be mistaken for members of the related goosefoot family (Chenopodiaceae) or amaranth family (Amaranthaceae). Those families, however, lack papery stipules on their leaves and do not have 3-sided ovaries.

STATISTICS 1,100 species widespread throughout the world and richly represented in the northern hemisphere, with a sprinkling of species in tropical areas. Many favor high mountains and cold climates, but arid western North America is rich in species that belong to the subfamily Eriogonoideae. Among the edible species are *Coccoloba* (tropical sea-grapes), rhubarb (*Rheum*), with its edible leaf stalks, French sorrel (*Rumex acetosa*), and the grain called buckwheat (*Fagopyrum esculentum*). Garden ornamentals include various species of lady's thumb (*Polygonum*) and our native wild buckwheats (*Eriogonum* spp.).

CALIFORNIA GENERA AND SPECIES

The region has 17 native genera and four nonnative or partly nonnative genera.

*POLYGONOIDEAE (KNOTWEED SUBFAMILY)*
Papery, sheathlike stipules on the leaves.

*Polygonum* (tear-thumb, bistort, knotweed, and others) is a large genus of annuals and herbaceous perennials with flowers in axillary clusters or terminal panicles and 4 to 6 white, green, pink, or red tepals.

*P. davisae* (Davis's knotweed) is a common perennial in the mountains with bright red new shoots and axillary green flowers.

*P. bistortoides* (bistort) is a common perennial in mountain meadows with spikes of white flowers.

*P. amphibium* (water knotweed) is widespread in ponds and marshes and features floating leaves and pink to red flowers.

*P. paronychia* (dune knotweed; see fig. a) is a creeping woody perennial from coastal sand dunes, with narrow leaves and white flowers.

*P. phytolaccaefolium* is an herbaceous perennial to 6 feet tall, with panicles of white flowers; it frequents mountain meadows.

*Oxyria digyna* (mountain sorrel) is a high-mountain perennial with kidney-shaped leaves, reddish flowers, and a 2-sided, rather than 3-sided ovary.

*Rumex* (docks) are rank annuals and perennials, some of them invasive nonnative weeds. They have elliptical leaves and greenish to reddish flowers usually with 6 tepals. The outer 3 are enlarged into wings for seed dispersal on wind. Native species include:

*R. californicus* (California dock), a stout perennial from marshes with 6-foot-tall flowering stalks.

*R. salicifolius* (willow-leaf dock), a creeping perennial from sand dunes and coastal marshes.

*R. hymenosepalus* (Indian rhubarb or canaigre; see fig. b), a giant desert dune perennial with rhubarblike leaves and conspicuous pink fruits.

*ERIOGONOIDEAE SUBFAMILY*
These characteristics pertain to Eriogonoideae of western North America.

LEAVES WITHOUT STIPULES AND FLOWERS USUALLY INSIDE INVOLUCRES *Pterostegia drymarioides* (fan-leaf) is a sprawling annual found in rocky foothills, with pairs of 2-lobed, fanlike leaves and minute green flowers.

*Chorizanthe* (spineflowers) are annuals with cylindrical or bell-shaped, spine-tipped involucres containing 1 flower each (the involucres resemble sepals and the colored tepals look like petals). Many species live in sandy habitats or in deserts.

*Eriogonum* (wild buckwheat) has dozens of species and varieties, and ranges from tiny annuals to shrubs, all on sunny sites in rocky or sandy soils. Numerous species occur in deserts and mountains. The species keys are based on details of the involucres and tepals. The *Jepson Manual* sorts them into the following four categories.

ANNUALS WITH UPRIGHT INVOLUCRES OR INVOLUCRES WITHOUT A STALK *Eriogonum luteolum* (rose buckwheat) is a diminutive, widespread annual with pale yellow or pink flowers.

ANNUALS WITH HANGING INVOLUCRES OR INVOLUCRES BORNE ON STALKS *Eriogonum spergulinum* is an airy plant with open panicles of white flowers and is widespread in open mountain forests.

*E. inflatum* (desert trumpet) (often) has conspicuously inflated, trumpetlike stems and tiny yellow star-shaped flowers.

PERENNIALS WITH JOINTED PERIANTH BASES in these species, the bottom of the perianth is slender and pinched in where it connects to the flowering stalk.

*Eriogonum umbellatum* (sulfur buckwheat; see fig. c) is a widespread cushionlike plant in rocky habitats, with umbel-like clusters of bright yellow flowers.

*E. compositum* (hot rock buckwheat) is a stout plant from northern California with large leaves and tall flowering stalks with rounded masses of white to pale yellow flowers.

*E. lobbii* (Lobb's buckwheat) is an alpine plant with matted leaves and pale yellow to pinkish flowers on stalks that are splayed close to the ground.

PERENNIALS AND SHRUBS WITH UNJOINTED PERIANTH BASES (THE LARGEST GROUP) In these species, the bottom of the perianth gradually tapers to the flowering stalk.

*Eriogonum nudum* (naked stem buckwheat; see fig. d) is widespread in rocky habitats throughout the state. It features a basal rosette of spoon-shaped leaves, naked flowering stalks, and ball-shaped clusters of white, pink, or yellow flowers.

*E. fasciculatum* (California buckwheat; see fig. e) is a small shrub with clusters of narrow leaves and flat-topped clusters of white or pale pink flowers. It is widespread throughout central and southern California.

## POLYGONOIDEAE (KNOTWEED SUBFAMILY)

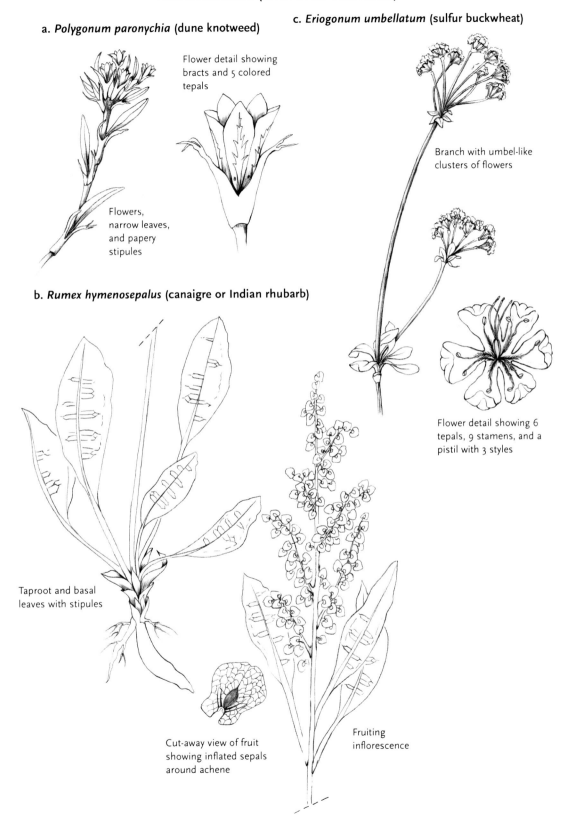

a. *Polygonum paronychia* (dune knotweed) — Flower detail showing bracts and 5 colored tepals; Flowers, narrow leaves, and papery stipules

b. *Rumex hymenosepalus* (canaigre or Indian rhubarb) — Taproot and basal leaves with stipules; Cut-away view of fruit showing inflated sepals around achene; Fruiting inflorescence

c. *Eriogonum umbellatum* (sulfur buckwheat) — Branch with umbel-like clusters of flowers; Flower detail showing 6 tepals, 9 stamens, and a pistil with 3 styles

POLYGONACEAE (BUCKWHEAT FAMILY)

# POLYGONACEAE (Buckwheat Family)

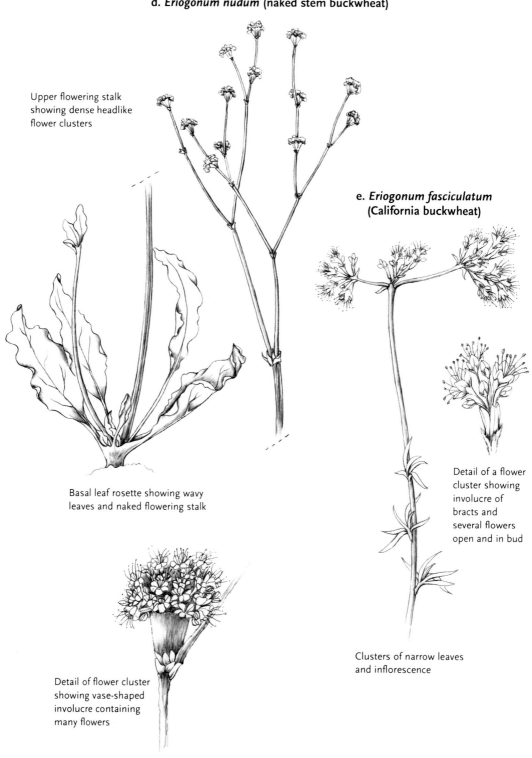

d. *Eriogonum nudum* (naked stem buckwheat)

Upper flowering stalk showing dense headlike flower clusters

Basal leaf rosette showing wavy leaves and naked flowering stalk

Detail of flower cluster showing vase-shaped involucre containing many flowers

e. *Eriogonum fasciculatum* (California buckwheat)

Detail of a flower cluster showing involucre of bracts and several flowers open and in bud

Clusters of narrow leaves and inflorescence

# PORTULACACEAE (Portulaca or Purslane Family)

RECOGNITION AT A GLANCE  Herbaceous plants with simple, fleshy leaves and flowers (often) with 2 sepals and 5 petals—an unusual combination.

VEGETATIVE FEATURES  Herbaceous annuals and perennials with simple, fleshy, alternate, basal, or opposite leaves.

FLOWERS  Small, mostly bisexual, wide open, sometimes showy, and solitary or in racemes and spikes.

FLOWER PARTS  Often 2 separate sepals (sometimes several), 5 separate petals (sometimes several), 5 stamens (sometimes numerous), and a single pistil with a (usually) superior ovary.

FRUITS  Capsules with shiny, black seeds.

RELATED OR SIMILAR-LOOKING FAMILIES  This leaf succulent family would probably not be confused with the other leaf succulent family, Crassulaceae, because of differences in flower parts and shapes: Crassulaceae has star-, bell-, or vase-shaped flowers with 5 sepals and 5 petals, and usually 5 partly separate pistils. The flowers of some "portulacs," such as the lewisias, have a design reminiscent of cactus blossoms, and currently botanists consider the two families closely related.

STATISTICS  400 species, of wide distribution but most prolific in North America, Australia, and South Africa. Many inhabit rocky or sandy soils. Some are eaten locally as potherbs and greens including our own miner's lettuce; none are major food crops. Ornamentals include lewisias for rock gardens and portulacas as bedding plants.

## CALIFORNIA GENERA AND SPECIES

The region has six mostly native genera.

*Portulaca* (purslane, portulaca) has flowers with a partially inferior ovary.

*P. oleracea* (purslane) is a sprawling garden weed with edible leaves and small yellow flowers.

*Calyptridium* (pussy paws) are taprooted annuals and perennials with a basal rosette of spoon-shaped leaves, densely massed flowers, 2 winged sepals, and 5 small petals.

*C. umbellatum* is widespread on sandy flats in the mountains and has headlike clusters of pink flowers that superficially resemble some eriogonums (wild buckwheats).

*Calandrinia* (red maids) are semisprawling annuals with (usually) magenta flowers.

*C. ciliata* is common in foothill grasslands.

*Claytonia* (spring beauty and others) are tap- or fibrous-rooted annuals and perennials that start with basal leaves, then produce flowering stems with a single pair of opposite, sometimes fused stem leaves. The racemes of small flowers are white or pink.

*C. perfoliata* (miner's lettuce; see fig. a) is an abundant edible annual that favors foothill woodlands. Its upper leaves form a disc around the stems.

*C. sibirica* (candy-stripe) is a short-lived perennial from moist forests, with separate stem leaves and pink flowers with dark pink stripes.

*C. gypsophiloides* (spring beauty) is an ephemeral annual on rocky soils with narrow, bluish green leaves and small, fragrant, pale pink flowers.

*Montia* are mostly perennials that feature 4 or more alternate or opposite stem leaves. Formerly, several of the claytonias were classified as montias.

*M. parvifolia* (little-leaf montia) is a short-lived perennial on shaded, mossy rocks, with sprawling stems and small white flowers.

*Lewisia* (lewisias) are perennials with basal rosettes of thick leaves and cymes of small to showy, cactuslike blossoms with several sepals, several petals, and an indefinite number of stamens. Many are considered choice rock garden plants.

*L. rediviva* (bitterroot; see fig. b) has narrow, cylinder-shaped leaves and single, large, cactuslike pink or white blossoms of great beauty. It lives on rock outcrops in the foothills and middle elevations of the mountains.

*L. cotyledon* (Siskiyou lewisia) has spoon-shaped leaves and open cymes of cactuslike pink, red, or rose-colored flowers. It favors rocky cliffs in the Klamath Mountains.

## PORTULACACEAE (Portulaca or Purslane Family)

### a. *Claytonia perfoliata* (miner's lettuce)

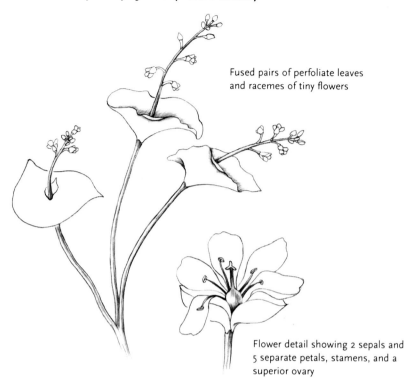

Fused pairs of perfoliate leaves and racemes of tiny flowers

Flower detail showing 2 sepals and 5 separate petals, stamens, and a superior ovary

### b. *Lewisia rediviva* (bitterroot)

Habit with fleshy basal leaves and large, multi-petaled, cactuslike flowers

## PRIMULACEAE (Primrose Family)

**RECOGNITION AT A GLANCE** Herbaceous plants often with simple basal or opposite leaves, and (often) umbels of flowers with 5 petals joined to form a tube and 5 stamens opposite (rather than alternate with) the petals.

**VEGETATIVE FEATURES** Herbaceous annuals and perennials, sometimes from fleshy roots or corms, with simple, often opposite, whorled, or basal leaves.

**FLOWERS** Small to showy, star-shaped, primroselike, or with reflexed petals; solitary, in umbels, or in leaf axils.

**FLOWER PARTS** 4 or 5 partly fused sepals, 4 or 5 petals joined to form a tube (sometimes very short), 4 or 5 stamens joined to the tube, and a single pistil with a superior, 1-chambered ovary.

**FRUITS** Capsules with a central stalk bearing several to many seeds (free-central placentation).

**RELATED OR SIMILAR-LOOKING FAMILIES** The primrose family should never be confused with the unrelated evening-primrose family (Onagraceae), which features flower parts in fours and an inferior ovary. The primrose family is similar in many respects to the pink family (Caryophyllaceae), but that family has no petal tube, and the stamens are usually alternate with the petals. Both families share a capsule-type fruit with free-central placentation.

**STATISTICS** 600 species, best represented in the temperate zone of the northern hemisphere. The family is especially diverse in the Himalayas and mountains of China. Popular garden ornamentals include cyclamen (*Cyclamen* spp.) and primroses (*Primula* spp.). None are used for food.

### CALIFORNIA GENERA AND SPECIES

The region has eight mostly native genera and one nonnative genus.

*Anagallis arvensis* (scarlet pimpernel) is a common, introduced garden weed with sprawling stems, opposite leaves, and small orange flowers with a central purple spot.

*Trientalis latifolia* is a tuberous perennial from shaded conifer forests with a single whorl of several broad leaves, threadlike flowering stalks above the leaves, and star-shaped white or pink flowers of 5 to 8 petals.

*Dodecatheon* (shooting star) are perennials with basal, elliptical to spoon-shaped leaves and umbels of pink, white, or rose-purple flowers with swept-back, cyclamenlike petals and stamens that "shoot" forward.

*D. clevelandii* (padre's shooting stars) is common in foothill meadows and woodlands and has somewhat "chunkier" flowers than the next species.

*D. hendersonii* (Henderson's shooting stars; see fig. a) is common in woodlands throughout the foothills and has similar but slimmer flowers.

*Androsace* (rock-jasmines) are seldom-noticed annuals in rocky mountain habitats with umbels of tiny, white flowers.

*Primula suffrutescens* (Sierra primrose; see fig. b) is unusual in its genus, being a sprawling, woody perennial with umbels of vivid magenta flowers and a yellow-green eye. It is found on rock scree near timberline.

*Centunculus minimus* (chaffweed) is a vernal pool annual with usually alternate leaves and axillary flowers with petals shorter than the sepals.

*Glaux maritima* (sea-milkwort) lives in salty habitats, has fleshy leaves, and bears tiny flowers with purple sepals and no petals.

*Lysimachia* (loosestrifes) are perennials with whorled leaves and axillary, star-shaped yellow blossoms. *L. nummularia* is a creeping, nonnative plant; *L. thyrsiflora* has upright stems and is native to wet habitats in the northern mountains.

## PRIMULACEAE (Primrose Family)

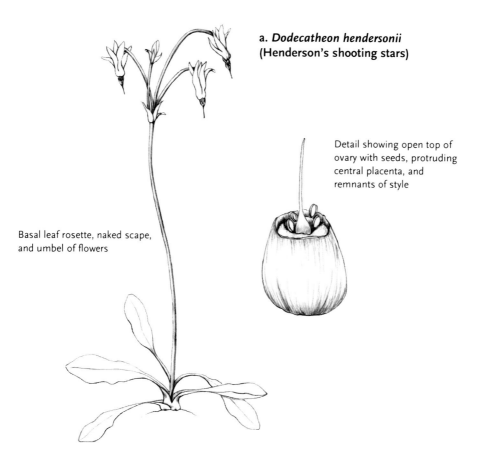

**a. *Dodecatheon hendersonii* (Henderson's shooting stars)**

Basal leaf rosette, naked scape, and umbel of flowers

Detail showing open top of ovary with seeds, protruding central placenta, and remnants of style

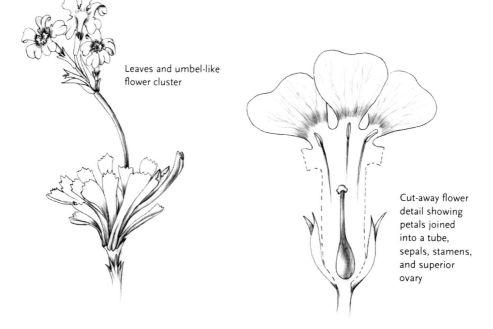

**b. *Primula suffrutescens* (Sierra primrose)**

Leaves and umbel-like flower cluster

Cut-away flower detail showing petals joined into a tube, sepals, stamens, and superior ovary

## PTERIDACEAE (Brake Fern Family)

**RECOGNITION AT A GLANCE** Mostly rock ferns with tough, wiry rachises (frond stems), pinnately to many times compound fronds (leaves), and (usually) sori (spore clusters) arranged along the underside edge of each frond segment.

**VEGETATIVE FEATURES** Small to medium-sized ferns from a usually short, vertical, underground rhizome; fronds divided into many small segments (pinnules) attached to a tough, wiry rachis and stipe (frond stems).

**FLOWERS** Missing; ferns reproduce by microscopic spores borne in spore sacs (sporangia) arranged into sori (brown patches) on the backside of the fronds.

**SORI** Each sorus consists of several to numerous tiny spore sacs that (mostly) follow the under edge of each pinnule. The edges of pinnules curl under (false indusium) to protect the young sori until spores mature and are ready to be shed.

**FRUITS** Mature, microscopic brown or black spores are released under pressure from their spore sacs into the air. Wind moves the spores to new homes.

**RELATED OR SIMILAR-LOOKING FAMILIES** Because of reproduction by spores from sori, ferns really are not likely to be confused with seed-bearing plant families. All the other families described in this book, including the conifers, bear seeds. But fern specialists have created many different fern families, several of which are native to California. Other fern families include Dryopteridaceae (wood fern family), Dennstaediaceae (bracken family), Aspleniaceae (spleenwort family), Polypodiaceae (polypody family), and Blechnaceae (blechnum fern family). For the amateur, distinguishing these families may prove difficult but the Pteridaceae has the distinctive combination of (usually) marginal sori with false indusia, and wiry frond stems. Most species grow in rocky places, where their roots are protected by tunneling under rocks.

**STATISTICS** 500 species found worldwide, mostly in dry, rocky habitats. Some of these ferns are grown as ornamentals in gardens, especially the genus *Adiantum* (maidenhair fern).

The stipes of some adiantums were used to create designs in California Indian baskets.

## CALIFORNIA GENERA AND SPECIES

The region has 10 genera, 27 native species, and three nonnative, occasionally naturalized species.

*FRONDS WITH FRAGILE, THIN PINNAE*

*Adiantum* (maidenhair ferns) have flag- or crescent-shaped pinnules on a polished black stipe and rachis. The sori are not continuous but separated into discrete structures along the frond's edge.

*A. aleuticum* (five-finger fern; see fig. a) has fronds with fingerlike divisions and grows along streams and seeps in the wooded parts of California.

*A. jordanii* (California maidenhair) lacks the fingerlike divisions of *aleuticum* and grows in dry woods throughout the foothills.

*A. capillus-veneris* (southern maidenhair) is similar to California maidenhair but grows by seeps and oases in central and southern California.

*Pentagramma triangularis* (silver- and goldback fern) has broadly triangular fronds and is immediately recognized by the underside being covered with silvery or yellowish wax. The sori follow veins. These ferns are common throughout the foothills.

*Cheilanthes* (bead ferns) are small and usually have a narrowly triangular frond shape. The pinnules look like tiny, rounded beads and are covered with scales or hairs underneath. The several species occur in rocky habitats throughout the state and are often difficult to tell apart.

*Pellaea* (cliffbrakes) have especially tough frond stems, and feature much divided fronds with oval to elliptical pinnules. They have no hairs underneath. The sori are continuous along the frond margins. Of the several species, the most widespread are:

*P. andromedifolia* (coffee fern; see fig. b). Fronds are pale to bluish green, turning brown to purple late in the year. Each pinna is an ellipse. Common throughout in the foothills, often in partial shade.

*P. mucronata* (birdsfoot fern). Fronds are grayish to bluish green and more finely divided than coffee fern. Each tiny pinnule ends in a minute, spinelike tip (use hand lens). Widespread in dry mountains throughout the state.

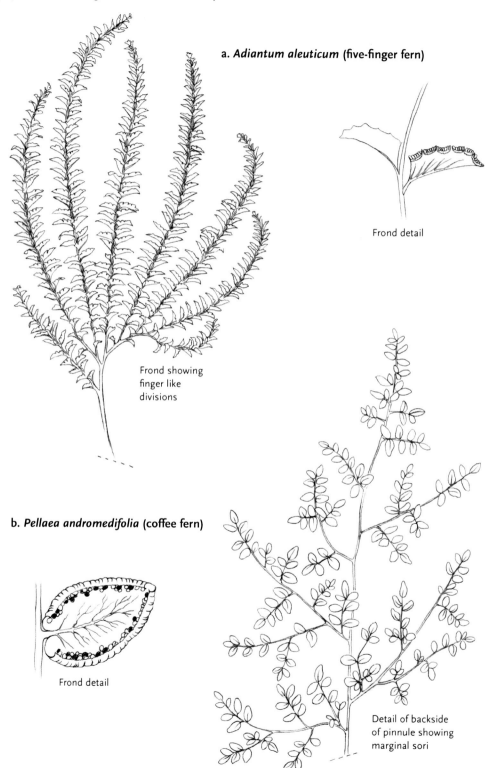

**a. *Adiantum aleuticum* (five-finger fern)**

Frond detail

Frond showing finger like divisions

**b. *Pellaea andromedifolia* (coffee fern)**

Frond detail

Detail of backside of pinnule showing marginal sori

# RANUNCULACEAE (Buttercup or Crowfoot Family)

RECOGNITION AT A GLANCE Herbaceous plants (often) with palmately lobed or ternately compound leaves and flowers with numerous, spirally arranged stamens and several separate pistils.

VEGETATIVE FEATURES Herbaceous annuals, perennials, and semiwoody vines with basal, alternate, or opposite leaves. Leaves are often palmately lobed, trifoliate, or ternately compound.

FLOWERS Often showy and of varied shapes—some are regular, others irregular, some are wind-pollinated, others are colorful and insect-pollinated. The flowers are single, in spikes, or in racemes.

FLOWER PARTS (Usually) 5 separate sepals and petals (petals sometimes missing), numerous spirally arranged stamens, and several to many separate, simple pistils with a superior ovary.

FRUITS (Usually) 1-seeded achenes or many seeded follicles.

RELATED OR SIMILAR-LOOKING FAMILIES The complex leaf patterns of some species are suggestive of certain members of the barberry family (Berberidaceae), but that family has a floral plan with 3 series of tepals in threes, and a fixed number of stamens. The flowers of several genera resemble herbaceous members of the rose family (Rosaceae), but that family often has stipules on its leaves and a row of sepal-like bracts outside the sepals. Neither of these features is found in the buttercup family.

STATISTICS 1,700 species of worldwide distribution, especially diverse in the mountains of the northern hemisphere. California, the Mediterranean region, the Alps, and the mountains of China are rich in species. Many bloom early. The family includes highly toxic genera such as larkspur (*Delphinium* spp.) and monkshood (*Aconitum* spp.), some of which have been used medicinally in tiny doses. The many garden ornamentals include anemones, clematis, ranunculus, larkspurs, columbines, and globe flower (*Trollius* spp.).

## CALIFORNIA GENERA AND SPECIES

The region has 14 native and two nonnative genera.

### GENERA WITH REGULAR FLOWERS

GENERA WITH SEPALS AND PETALS *Ranunculus* (buttercups) usually have shiny yellow petals (a few have white or pink-tinted flowers).

*R. californicus* (California buttercup, fig. a) is common in foothill woodlands and has 10 or more petals.

*R. aquatilis* (water buttercup) is a mostly submerged pond plant with filmy leaves and white flowers.

Several species are nonnative weeds.

*Aquilegia* (columbines) are perennials with ternately compound leaves and spurred petals that alternate with colored sepals.

*A. formosa* (red columbine; see fig. b) lives in forests and has hanging red and yellow flowers.

*A. pubescens* (alpine columbine) lives in rock scree near timberline and has horizontal, pale yellow, pink, or white flowers with very long spurs.

GENERA WITH SEPALS ONLY; SEPALS COLORED AND PETAL-LIKE *Anemone* (wind flowers) are perennials with white, purple, or pinkish sepals and numerous pistils that ripen into achenes.

*A. occidentalis* (western pasque flower) lives in high mountains and has spectacular heads of achenes topped with feathery styles.

*Isopyrum* (rue-anemones) are delicate woodland plants with ternately compound leaves; white, anemonelike flowers; and follicles.

*Clematis* (clematis; virgin's bower) are semiwoody vines with compound leaves and starry, white or cream-colored flowers.

GENERA WITH SEPALS ONLY; SEPALS GREEN *Thalictrum* (meadowrues) have ternately compound leaves and (often) unisexual, greenish flowers. They live in moist woodlands and mountain meadows.

*Myosurus* (mousetails) are tiny annuals with stamens on conelike receptacles, and spurred sepals.

*Actaea arguta* (baneberry) has large, compound leaves and spikelike clusters of tiny, petalless, whitish flowers followed by red, waxy, poisonous berries. It lives in moist forests.

GENERA WITH IRREGULAR FLOWERS THAT HAVE BOTH PETALS AND SEPALS

*Aconitum columbianum* (monkshood) is a tall, mountain-meadow perennial with racemes of hooded blue or purple flowers. The petals are modified into nectar-bearing spurs inside the hooded, upper sepal.

*Delphinium* (larkspurs) are tap- or tuberous-rooted perennials with spikelike racemes of blue, purple, white, pink, yellow, or red flowers. The upper petal-like sepal forms a slender pointed spur; the 4 or 5 petals form a 2-lipped design inside the sepals. Keys often require knowledge of the roots in order to identify species.

*D. cardinale* and *D. nudicaule* (scarlet larkspurs) have scarlet or red flowers.

*D. gypsophilum* and *D. hesperium* var. *pallescens* have pale, near white flowers.

*D. luteum* (yellow larkspur) has pale yellow flowers.

*D. purpusii* has pink-purple flowers.

### a. *Ranunculus californicus* (California buttercup)

### b. *Aquilegia formosa* (red columbine)

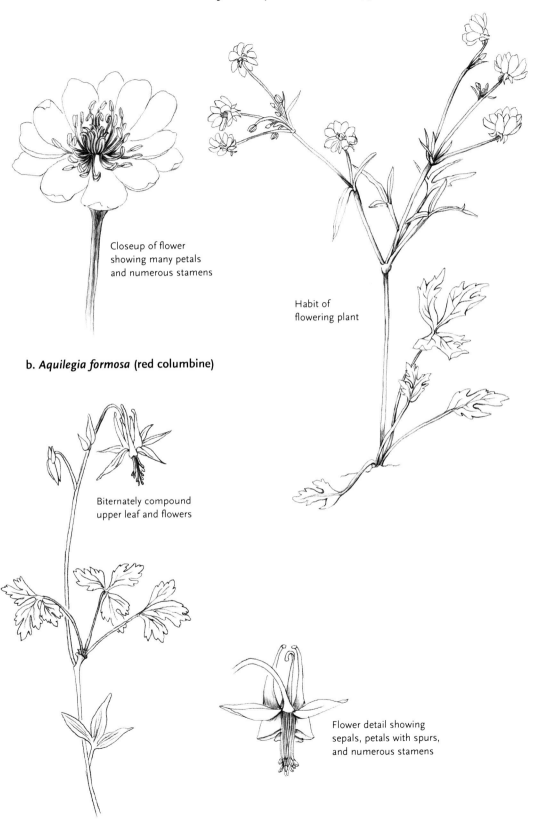

Closeup of flower showing many petals and numerous stamens

Habit of flowering plant

Biternately compound upper leaf and flowers

Flower detail showing sepals, petals with spurs, and numerous stamens

RANUNCULACEAE (BUTTERCUP OR CROWFOOT FAMILY)

# RHAMNACEAE (Buckthorn Family)

RECOGNITION AT A GLANCE Woody plants with simple leaves, a pinnate-arcuate vein pattern, numerous small flowers in dense cymes and panicles, and a 3-sided ovary.

VEGETATIVE FEATURES Evergreen and deciduous shrubs and small trees with simple leaves, stipules, and a pinnate-arcuate vein pattern.

FLOWERS Usually bisexual, greenish or colored; tiny, numerous, and in umbel-like cymes or dense panicles.

FLOWER PARTS 4 or 5 separate sepals, 4 or 5 separate, sometimes hooded petals (sometimes missing), 4 or 5 stamens attached to a disclike hypanthium, and a single pistil with a superior (or partly inferior), 3-sided, 3-chambered ovary.

FRUITS Fleshy drupes or capsules with a few large seeds.

RELATED OR SIMILAR-LOOKING FAMILIES Other woody families with numerous small flowers include the sumac family (Anacardiaceae), but that family never has blue or purple flowers, and the leaves do not have a pinnate-arcuate vein pattern. Also, most buckthorn family leaves are not resinously scented (a few *Ceanothus* have fragrant leaves).

STATISTICS Around 900 species, diverse in the tropics but also well represented in arid habitats in California, Mexico, the Mediterranean region, and Australia. The fruits of Chinese jujube (*Ziziphus jujuba*) are sometimes grown for food. Several ornamental shrubs include the buckthorns (*Rhamnus* spp.) and wild lilacs (*Ceanothus* spp.).

## CALIFORNIA GENERA AND SPECIES

The region has six native genera.

*Ceanothus* spp. (wild lilacs; deerbrush; buckbrush) have many species ranging from deciduous or evergreen, woody ground covers to small trees. All feature fragrant white, pink, purple, or blue flowers with colored sepals and petals. The shrubs are common components of foothill woodlands and chaparral. The genus is divided into two subgenera:

*Ceanothus* (see figs. a and b) are shrubs with alternate leaves, deciduous stipules, and capsules without special horns or knobs.

*Cerastes* are shrubs with (usually) opposite, often hollylike leaves, thick corky stipules, and capsules with horns or knobs on top.

*Rhamnus* (buckthorn, coffee berry) are deciduous or evergreen shrubs with tiny, yellowish green, star-shaped flowers and 2- or 3-seeded red or purple berrylike drupes.

*R. californica* (coffee berry; see fig. c) is evergreen, has leathery, elliptical leaves, flowers with petals, and dark purple fruits.

*R. purshianus* (cascara sagrada) is large and deciduous and has thin, elliptical leaves, petalless flowers, and dark purple fruits.

*R. ilicifolia* (holly-leaf redberry; see fig. d) is dioecious and has evergreen, hollylike leaves, petalless flowers, and red fruits.

Other shrubby genera are mainly in the deserts.

### a. *Ceanothus integerrimus* (deerbrush)

### b. *Ceanothus cuneatus* (buckbrush)

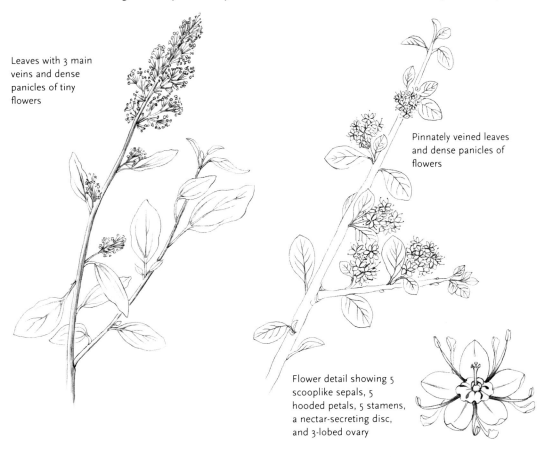

Leaves with 3 main veins and dense panicles of tiny flowers

Pinnately veined leaves and dense panicles of flowers

Flower detail showing 5 scooplike sepals, 5 hooded petals, 5 stamens, a nectar-secreting disc, and 3-lobed ovary

### c. *Rhamnus californica* (coffee berry)

Cut-away flower detail showing 5 sepals, longer petals, and 5 stamens

### d. *Rhamnus ilicifolia* (holly-leaf redberry)

Leafy branch

RHAMNACEAE (BUCKTHORN FAMILY) 151

# ROSACEAE (Rose Family)

**RECOGNITION AT A GLANCE** Plants of highly variable habits and leaf designs. Flowers resemble single roses with 5 separate sepals, petals, and numerous stamens attached to a shallow cuplike or bowl-shaped hypanthium.

**VEGETATIVE FEATURES** Herbaceous annuals, perennials, shrubs, and small trees with (usually) alternate leaves that often bear stipules. Leaves may be simple, highly dissected, compound, entire, or toothed.

**FLOWERS** Red, pink, yellow, or white; tiny to large; resembling single roses; arranged in varied ways.

**FLOWER PARTS** Mostly 5 separate sepals and petals (sometimes missing) and numerous (usually) stamens attached to a saucer- or bowl-shaped hypanthium. Pistils may be single with a superior to inferior ovary, multiple on a flat receptacle, or borne inside a hypanthium.

**FRUITS** The many different kinds include achenes, follicles, drupes, accessory fruits, aggregate fruits, and pomes. The family is noted for lacking true capsules and berries.

**RELATED OR SIMILAR-LOOKING FAMILIES** The flower design is reminiscent of some members of the buttercup family (Ranunculaceae) but the herbaceous roses have stipules on their leaves and a row of sepal-like bracts outside the actual sepals. Some of the species with small flowers also have a design like certain saxifrages (Saxifragaceae), but saxifrage flowers lack sepal-like bracts and generally have 2 separate to partly joined pistils.

**STATISTICS** A prominent family of 3,000 species found throughout the world but with the greatest diversity in the northern hemisphere. It is noted for its wide variety of edible fruits including apples, pears, peaches, plums, cherries, almonds, strawberries, raspberries, and more. The rose, arguably the world's most popular flower, is a phenomenon in gardens and for the cut-flower trade, with hundreds of named cultivars. Many other ornamentals for gardens include a large array of shrubs and flowering trees.

## CALIFORNIA GENERA AND SPECIES

The region has 26 native or partly native genera and three nonnative genera.

### HERBACEOUS GENERA

**PLANTS WITH TRIFOLIATE LEAVES** *Fragaria* (strawberry) has broad leaflets, white flowers, and berrylike accessory fruits.

*Duchesnea indica* (mock strawberry) is an introduced herb with a similar appearance to the strawberry, but with yellow flowers and tasteless fruits.

*Sibbaldia procumbens* is a sprawling mountain plant with small leaves and tiny, light yellow flowers.

**PLANTS WITH PINNATELY OR PALMATELY COMPOUND LEAVES** *Potentilla* (cinquefoil) has palmately or pinnately compound leaves; red, yellow, or whitish flowers with numerous stamens; and achenes with styles that fall off when ripe.

*P. glandulosa* (sticky cinquefoil; see fig. a) has many races from near sea level to above timberline.

*P. fruticosa* (shrubby cinquefoil) is exceptional in being a small shrub. It grows in the high mountains.

*Horkelia* (horkelia) has pinnately compound, often fragrant leaves and small, clawed, white or pinkish flowers.

*Ivesia* (ivesia; mouse-tails) has highly dissected leaves and small, yellow, pink, or white flowers with unclawed petals. Most species live in the high mountains.

*Geum* (avens) has pinnately compound leaves and yellow or cream-and-pink flowers followed by achenes with plumed or barbed styles.

### SHRUBBY GENERA

These genera have many small flowers in dense clusters and dry fruits (achenes and follicles).

*Lyonothamnus floribundus* (island ironwood) is a small, suckering tree with opposite leaves and flat-topped clusters of white flowers.

*Physocarpus* (ninebark) is a shrub with alternate, palmately lobed leaves and rounded, headlike clusters of white flowers.

*Adenostoma* (chamise; red-shanks) are large shrubs with narrow, needlelike leaves and dense panicles of white flowers.

*A. fasciculatum* (chamise) is among our most common chaparral shrubs.

*Aruncus vulgaris* (goatsbeard) is a woody, rhizomatous perennial with large, fernlike leaves and dense plumes of minute white flowers.

*Purshia* (cliff-rose; antelope brush) are large shrubs with 3-toothed leaves and white or yellow, roselike flowers.

*Petrophyton caespitosum* (rock-spiraea) forms dense mats with narrow, simple leaves and candlelike spikes of tiny white flowers.

*Spiraea* (spiraea; steeple bush) are rhizomatous shrubs with toothed, deciduous leaves, and dense cymes or spikelike panicles of rose-pink flowers.

*Holodiscus* (creambush; ocean spray) are deciduous shrubs with simple, soft, coarsely toothed leaves and dense panicles of tiny cream-colored flowers.

*Cercocarpus* (mountain-mahoganies; see fig. b) are large shrubs or small trees with simple, sometimes toothed leaves and clusters of cream-colored, petalless flowers followed by plumed fruits.

WOODY GENERA

GENERA WITH FLESHY FRUITS AND SUPERIOR OVARIES *Rubus* (black-, salmon-, thimble-, and raspberries) are creeping to upright, often prickly shrubs, with palmately lobed to compound leaves, white or pinkish flowers, and edible aggregate fruits. Some species are introduced and extremely aggressive.

*Prunus* (wild plums and cherries; see fig. c) are small to large, sometimes thorny shrubs with evergreen or deciduous leaves and racemes of white or pink blossoms and fleshy drupes, many of which are edible when cooked with sugar.

*Oemleria cerasiformis* (oso-berry) is a medium-sized, deciduous shrub with simple, entire leaves, racemes of white flowers, and dark purple drupes.

WOODY GENERA

GENERA WITH FLESHY FRUITS AND (APPARENTLY) INFERIOR OVARIES

*Rosa* (wild roses; see fig. d) are prickly shrubs with pinnately compound leaves and small clusters of fragrant, pink flowers followed by red hips (hypanthiums enclosing several achenelike fruits).

*Heteromeles arbutifolia* (toyon; California holly) is a large evergreen shrub or small tree with tough, toothed, simple leaves; rounded clusters of small white flowers; and bright red pomes.

*Malus fusca* (native crabapple) is a small deciduous tree with simple leaves and a mittenlike lobe; white, apple-blossomlike flowers; and small purplish pomes.

*Crataegus* (hawthorns) are small, thorny, deciduous shrubs or trees with leaves lined with coarse lobes and teeth; clusters of small, white flowers; and red-purple pomes.

*Amelanchier* (service berries) are large, deciduous shrubs with broadly elliptical leaves toothed on the upper half; clusters of white, apple-blossomlike flowers; and small, red-purple pomes.

## a. *Potentilla glandulosa* (sticky cinquefoil)

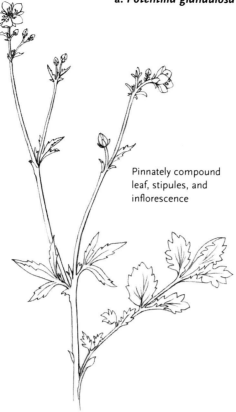

Pinnately compound leaf, stipules, and inflorescence

Flower detail showing multiple stamens and pistils

## b. *Cercocarpus betuloides* (mountain mahogany)

Leaves with characteristic pinnate vein pattern and cluster of flowers with tapered hypanthium, inferior ovary, and numerous stamens

Fruit detail showing long, plumelike style

### c. *Prunus ilicifolia* (holly-leaf cherry)

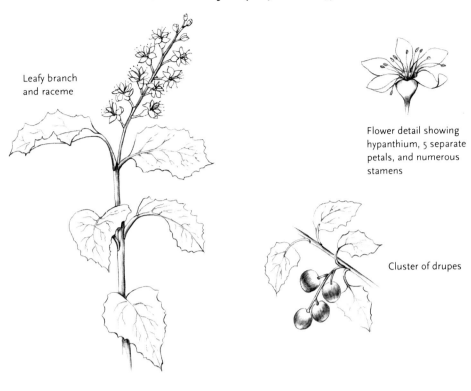

Leafy branch and raceme

Flower detail showing hypanthium, 5 separate petals, and numerous stamens

Cluster of drupes

### d. *Rosa californica* (California rose)

Pinnately compound leaf with stipules at the base and 2 flowers showing 5 separate petals and numerous stamens

Hypanthium with sepals on top

Detail of simple pistils inside hypanthium

# RUBIACEAE (Madder or Coffee Family)

RECOGNITION AT A GLANCE Plants with pairs or whorls of simple leaves often with stipules, and clusters of tiny, star-shaped flowers with inferior ovaries.

VEGETATIVE FEATURES Annual and perennial herbs and shrubs. Simple, entire, lance-shaped, ovate, or linear leaves in pairs with stipules, or in whorls; stems sometimes four-sided (square in cross-section).

FLOWERS Tiny to small; white, greenish yellow, or purple; and usually star-shaped (California species). Arranged in small clusters and heads.

FLOWER PARTS 4 or 5 minute (often apparently missing) sepals, 4 or 5 petals joined to form a short tube, 4 or 5 stamens, and a single pistil with an inferior, 2-lobed ovary.

FRUITS Split into 2 tiny, one-seeded nutlets often with pronged hairs on the outer surface; or they may sometimes be a berry or capsule.

RELATED OR SIMILAR-LOOKING FAMILIES Many features are shared between Rubiaceae and Caprifoliaceae (honeysuckle family). Besides the fact that most honeysuckles live in temperate climates and most Rubiaceae are tropical, the following features are helpful in separating the two families: California honeysuckles have bell-shaped, tubular, sometimes 2-lipped flowers, their leaves lack stipules, and their fruits are (mostly) fleshy and berrylike. California Rubiaceae have star-shaped, regular flowers; leaves often with stipules or arranged in whorls; and dry, nutlet-type fruits.

STATISTICS Prominent dicot family with around 6,000 species; most are tropical and range from herbaceous plants to rain forest trees and everything in between. Although few provide food, those essential coffee beans come from the African *Coffea arabica*, and cola flavoring is extracted from the seeds of the African *Cola nitida*. The many garden ornamentals include gardenias and clerodendrons.

## CALIFORNIA GENERA AND SPECIES

The region has four mostly native genera with 50 species and 2 nonnative genera and species.

### GENERA WITH WHORLED LEAVES

*Galium* (bedstraws; see fig. a) is a large genus that features sprawling to cushion-forming plants with woody to herbaceous stems; stems armed with minute, downward-pointing barbs; and white, purplish, or yellow-green flowers. Fruits are often important in identifying the species. A few are introduced, noxious weeds while several native species are rare and restricted in distribution. The species occur in many different habitats throughout the state.

*Sherardia arvensis* (field madder) resembles bedstraws but has pink-purple flowers. It comes from Europe.

### GENERA WITH OPPOSITE LEAVES

*Kelloggia galioides* is a widespread but little-noticed herbaceous perennial from montane forests with white to pale pink, trumpet-shaped flowers.

*Cephalanthus occidentalis* (western buttonbush or button-willow; see fig. b) is a large deciduous shrub of foothill and valley watercourses with ovate leaves and buttonlike heads of white flowers.

### a. *Galium* sp. (bedstraw)

Whorled leaves and tiny flowers

Cut-away flower detail showing hairs on inferior ovary, petals, stamens, and forked style

### b. *Cephalanthus occidentalis* (western buttonbush or button-willow)

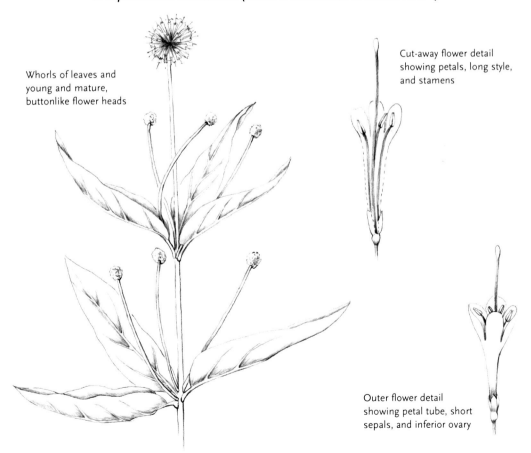

Whorls of leaves and young and mature, buttonlike flower heads

Cut-away flower detail showing petals, long style, and stamens

Outer flower detail showing petal tube, short sepals, and inferior ovary

RUBIACEAE (MADDER OR COFFEE FAMILY)

# RUTACEAE (Rue or Citrus Family)

**RECOGNITION AT A GLANCE** Shrubs and small trees with untoothed leaves dotted with oil glands that impart a strong, pungent, citrusy odor.

**VEGETATIVE FEATURES** Shrubs and small trees with alternate, simple to trifoliate, entire leaves covered with tiny, dark oil glands that impart a pungent odor.

**FLOWERS** Small, white or dark purple, star-shaped or erect and nearly closed, and borne in small clusters.

**FLOWER PARTS** 4 or 5 separate sepals and petals, 10 to 20 stamens, and a single pistil with a superior, lobed ovary. (The ovary often also has oil glands.)

**FRUITS** Varied—achenes (and samaras), capsules, drupes, and berries.

**RELATED OR SIMILAR-LOOKING FAMILIES** The rue family does not closely resemble other native families with the possible exception of the sumac family (Anacardiaceae), which is also noted for leaves with a pungent smell, and the production of many small, often whitish flowers. The sumacs, however, do not have the same oil glands embedded in their leaves, and their flowers are mostly carried in dense paniclelike sprays.

**STATISTICS** 1,500 species, mainly from the tropics; many occur in South Africa, Australia, and Southeast Asia. Economically, the most important genus is *Citrus* (orange, grapefruit, lemon, tangerine, lime). Several ornamentals are grown in mild climates including Mexican mock orange (*Choisya ternata*) and breath-of-heaven (*Coleonema* spp.).

## CALIFORNIA GENERA AND SPECIES

The region has three native genera and species.

*Cneoridium dumosum* (berry-rue) is a small shrub from the far south coast with simple, opposite leaves; white, star-shaped flowers; and red, berrylike fruits.

*Thamnosma montana* (turpentine broom) is a desert shrub with small, alternate leaves; dark purple flowers; and a 2-lobed capsule covered with glands.

*Ptelea crenulata* (hopbush; see fig.) is a large shrub from foothill canyons with deciduous, trifoliate leaves; white flowers; and circular samaras. It is particularly abundant in the Mt. Diablo region but is widely scattered elsewhere.

## *Ptelea crenulata* (hopbush)

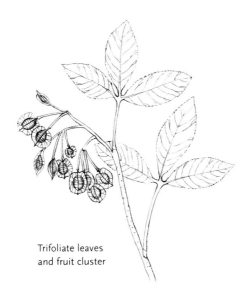

Trifoliate leaves and fruit cluster

Samara detail

Flower detail

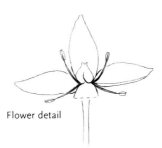

Flower raceme showing 4 petals, stamens, and single pistil

# SALICACEAE (Willow Family)

**RECOGNITION AT A GLANCE** Deciduous, dioecious shrubs and trees with simple leaves, petalless catkins of flowers, and hairy seeds.

**VEGETATIVE FEATURES** Deciduous, dioecious shrubs and trees with simple, unlobed leaves and pairs of stipules, which are often shed after leaves develop.

**FLOWERS** Tiny, unisexual, greenish, wind-pollinated, and in cylindrical spikes or catkins.

**FLOWER PARTS** Male flowers have a single bract with 1 or more stamens; female flowers have a single bract and a single pistil with a superior ovary.

**FRUITS** Capsules with numerous hairy, wind-dispersed seeds.

**RELATED OR SIMILAR-LOOKING FAMILIES** The willow family joins ranks with others that bear unisexual, petalless flowers in catkins. The garrya family (Garryaceae) features pairs of broad, evergreen leaves and hanging catkins; the birch family (Betulaceae) has male and female catkins on the same plant and conelike fruits or nuts. The oak family (Fagaceae) has male and female flowers on the same plant, an acorn in fruit, and female flowers in small axillary clusters, not in catkins. (The garrya family is not detailed in this book.)

**STATISTICS** 340 species distributed across the temperate northern hemisphere, and uncommon in the tropics. Both the cottonwoods (*Populus* spp.) and willows (*Salix* spp.) provide fast-growing trees and shrubs widely used for windbreaks and shade. Several willows are also ornamentals in gardens. Willow bark is a source of the active ingredient in aspirin, salicylic acid. Willow twigs are often bundled together and planted on steep banks as fast-growing soil retainers.

## CALIFORNIA GENERA AND SPECIES

The region has two mostly native genera and many species.

*Salix* (willows) are shrubs or trees mostly with lance-shaped or linear leaves and upright flower catkins. They favor riparian areas throughout the state, and the many species are often difficult to identify.

*S. arctica* and *S. nivea* (alpine and snow willows) are unusual in being prostrate, woody ground covers that live in the alpine regions of the high mountains.

*S. lasiolepis* (arroyo willow; see fig. a) is a small tree or large shrub common along watercourses in the foothills.

*Populus* (cottonwoods) are trees with triangular or rounded leaves and hanging catkins of flowers.

*P. fremontii* (Fremont cottonwood, see fig. b) is a common riparian tree with broad, scalloped leaves.

*P. balsamifera* var. *trichocarpa* (black cottonwood) is also a common riparian tree with lance-ovate, nearly entire leaves.

*P. tremuloides* (quaking aspen) is a clonal tree with whitish bark and round, scalloped leaves that flutter in breezes. It lives in the high mountains and turns luminous shades of gold and orange in the fall.

### a. *Salix lasiolepis* (arroyo willow)

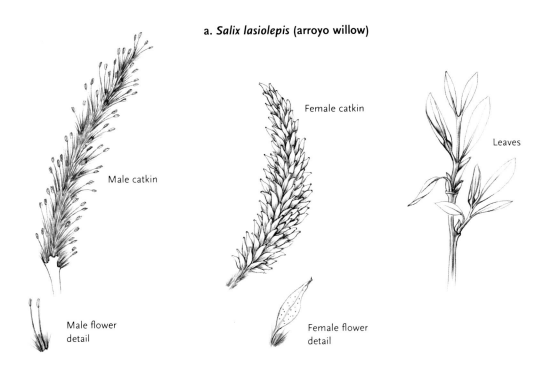

### b. *Populus fremontii* (Fremont cottonwood)

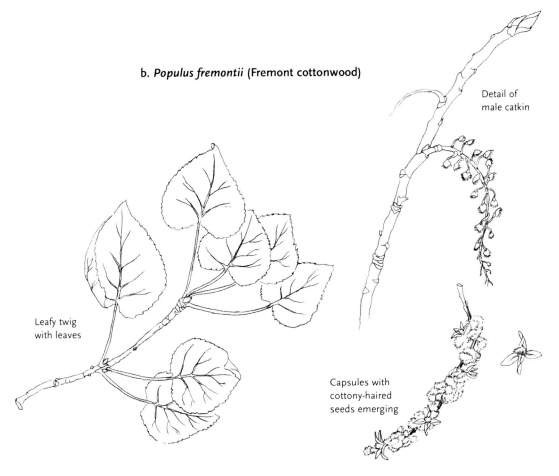

# SAXIFRAGACEAE (Saxifrage Family)

RECOGNITION AT A GLANCE Herbaceous plants (usually) with a basal rosette of broad, often rounded leaves; small, white, yellow, or pink flowers in racemes or panicles; and (usually) 2 separate or partly joined pistils.

VEGETATIVE FEATURES Mostly herbaceous perennials (usually) with a basal rosette of broad, often round and scalloped leaves, and (usually) no stipules.

FLOWERS Tiny, open to bell-shaped; white, yellow, or pink; and borne in panicles, cymes, or racemes.

FLOWER PARTS 4 or 5 separate sepals and petals, the same or twice the number of stamens, all joined to a cup-shaped or shallow hypanthium, and (usually) 2 pistils that are separate or partly joined.

FRUITS Capsules and follicles with many seeds.

RELATED OR SIMILAR-LOOKING FAMILIES The saxifrages might sometimes be confused with herbaceous members of the rose family but those have stipules on their leaves, a row of leaflike bracts under the sepals, and seldom 2 separate or partly joined pistils. At one time, the saxifrage family contained several woody plants, which have now been separated into different families such as the mock-orange family (Philadelphaceae), gooseberry family (Grossulariaceae), hydrangea family (Hydrangeaceae), and escallonia family (Escalloniaceae). (Only the gooseberry family is detailed in this book.)

STATISTICS Around 600 species, diverse in the mountainous parts of the northern hemisphere. Several saxifrages (*Saxifraga* spp.) are grown in rock gardens and several others are used in shade gardens (*Heuchera* spp.), or occasionally grown as house plants (e.g., piggyback plant, *Tolmiea menziesii*).

## CALIFORNIA GENERA AND SPECIES

The region has 16 mostly native genera.

GENERA WITH ROUND, USUALLY SCALLOPED OR COARSELY TOOTHED LEAVES

GENERA WITH UNLOBED OR UNFRINGED PETALS *Darmera peltata* (umbrella plant; Indian rhubarb) is a rhizomatous perennial that grows along mountain streams with 3-foot-broad, umbrellalike leaves and cymes of pale pink flowers.

*Jepsonia* (jepsonias) are corm-bearing perennials from rocky foothill forests with spring leaves and fall-blooming white flowers.

*Heuchera* (alumroots; see fig. a) are perennials from rootstocks with sometimes mottled leaves and airy panicles of white, pale yellow, or pink, bell-shaped flowers, and 5 stamens. They live in many rocky habitats.

*Tiarella trifoliata* var. *unifoliata* (sugar scoops) is a similar-looking perennial with short racemes of nodding, white, bell-shaped flowers, and 10 stamens. The fruit looks like an old-fashioned sugar scoop. The plant lives in moist coastal conifer forests.

*Bolandra californica* has similar leaves but the bell-shaped flowers are green and purple. It lives by streams in the central and northern Sierra.

*Boykinia* (brook saxifrages) have open panicles of tiny, upright, white flowers, and 5 stamens. They live on rocky banks by forested streams and waterfalls.

GENERA WITH FRINGED OR TOOTHED PETALS *Lithophragma* (woodland stars; see fig. b) have small, sometimes purple-tinted leaves and racemes of white or pale pink flowers with snowflakelike petals. They live in open forests and woodlands.

*Tellima grandiflora* (fringe-cups) is a perennial with racemes of small, bell-shaped, horizontally held flowers with fringed petals that start green and fade to rose color. It is widespread in coastal woodlands and forests.

*Mitella* (mitreworts; bishop's caps) are mostly stoloniferous perennials with racemes of small, usually green flowers with snowflake-like petals. They live in boggy forests and mountain meadows.

### GENERA WITH OTHER LEAF SHAPES

*Chrysosplenium glechomifolium* (golden saxifrage) is a sprawling herb with petalless flowers and yellow sepals, that lives in moist coastal conifer forests.

*Parnassia* (grass-of-Parnassus) are perennials with elliptical leaves and clusters of cream-colored flowers with veined petals, and jewel-like sterile stamens around 5 fertile stamens. They live in mountain meadows.

*Saxifraga* (saxifrages) have variable leaves and cymes of white or pink flowers with 10 stamens. They live on mossy banks in woodlands, along mountain streams, and in wet meadows.

# SAXIFRAGACEAE (Saxifrage Family)

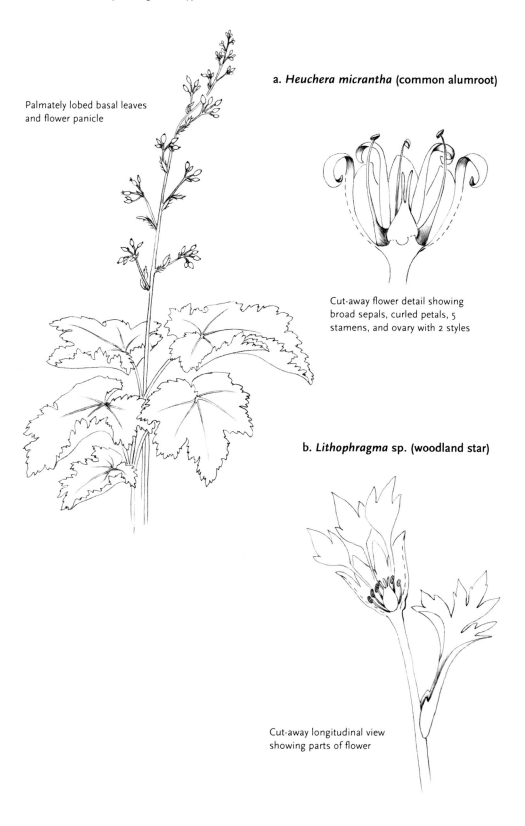

Palmately lobed basal leaves and flower panicle

**a. *Heuchera micrantha* (common alumroot)**

Cut-away flower detail showing broad sepals, curled petals, 5 stamens, and ovary with 2 styles

**b. *Lithophragma* sp. (woodland star)**

Cut-away longitudinal view showing parts of flower

# SCROPHULARIACEAE (Figwort or Snapdragon Family)

RECOGNITION AT A GLANCE (Mostly) herbaceous plants with 2-lipped flowers and 2-chambered capsules containing many small seeds.

VEGETATIVE FEATURES Herbaceous annuals, perennials, and small shrubs, often with opposite leaves. Leaves may be simple and entire, toothed, deeply lobed, or pinnately dissected. Some are root parasites despite having green leaves.

FLOWERS (Usually) irregular and 2-lipped, often showy, in a variety of colors, mostly in spikes, racemes, or panicles.

FLOWER PARTS (Usually) 5 partly joined sepals, 5 irregular, 2-lipped petal lobes joined to form a tube, 4 (sometimes 2 or 5) stamens joined to the tube, and a single pistil with a superior, 2-chambered ovary.

FRUITS Capsules with many small seeds.

RELATED OR SIMILAR-LOOKING FAMILIES The 2-lipped design of "scroph" flowers resembles flowers in several other families. The mints (Lamiaceae) have a similar design but usually have fragrant leaves and a deeply 4-lobed ovary. The acanths (Acanthaceae) have a similar design but the ovary only contains 4 large seeds expelled explosively from the capsule. The bignons (Bignoniaceae) have a similar design but are woody plants with long, beanlike seed pods and winged seeds. The broomrapes (Orobanchaceae) are parasites with no green chlorophyll. (Only the mint family is detailed in this book.)

Recent studies indicate that the Scrophulariaceae is not really a single, coherent family. A few of our species will probably be retained in the family, but many will be separated into the Veronicaceae or Plantaginaceae depending on how the names are used, while the monkeyflowers (*Mimulus* spp.) belong to their own family, Phrymaceae. Genera with a galea-type upper lip will be combined with the parasitic broomrape family (Orobanchaceae).

STATISTICS 3,000 species worldwide but few species live in tropical lowlands. The major diversity is in the Mediterranean region, California, and other arid areas. Foxglove (*Digitalis purpurea*) is an important source of digitalis, used to treat heart ailments. Many genera are ornamentals in gardens, including snapdragon (*Antirrhinum majus*), monkeyflowers (*Mimulus* spp.), lady's pocket book (*Calceolaria* spp.), hebe (*Hebe* spp.), veronicas (*Veronica* spp.), penstemons (*Penstemon* spp.), toadflaxes (*Linaria* spp.), and mulleins (*Verbascum* spp.).

## CALIFORNIA GENERA AND SPECIES

The region has 27 partly or mostly native genera and six nonnative genera.

### GENERA WITH NEARLY REGULAR, NOT 2-LIPPED FLOWERS

*Verbascum* (mulleins) are nonnative biennials with rosettes of large leaves and tall spikelike clusters of nearly regular, flat, yellow or white flowers.

*Veronica* (speedwells; see fig. a) are native and nonnative annuals and perennials with blue or purple flowers, 4 flat petals, and 2 stamens. Many live in wet places and meadows.

### GENERA WITH CLEARLY IRREGULAR, USUALLY 2-LIPPED FLOWERS

GENERA WITH THE UPPER LIP FORMING A TUBULAR, HOOKED, OR HATCHET-SHAPED GALEA ENFOLDING THE STAMENS *Parentucellia viscosa* and *Bellardia trixago* are Mediterranean weeds: the first has bright yellow flowers and the second has bicolored, purple and white flowers.

*Castilleja* (Indian paintbrush, owl's clover, and others) are annuals and perennials with variable leaves, and spikes of red, orange, yellow, rose, or pink flowers. Much of the color comes from showy bracts below the flowers. The paintbrushes are divided into annuals with colored petals and perennials mostly with green petals and colored sepals.

Annuals include the following species.

*C. exserta* and *C. densiflora* (owl's clover), two similar-looking grassland annuals with mops of rose-purple flowers.

*C. rubicundula* and varieties (cream sacs) are grassland annuals sporting spikes of flowers with inflated, white to yellow petals.

Perennials include the following species.

*C. foliolosa* (woolly paintbrush; see fig. b) is a bushy plant with woolly gray leaves that is found in chaparral.

*C. lemmonii* (Lemmon's paintbrush) has slender spikes of magenta flowers and lives in high mountain meadows.

*C. miniata* (meadow paintbrush) has tall spikes of red, pink, or red-orange flowers and lives in mountain meadows.

*C. applegatei* (wavy-leaf paintbrush) has sticky, wavy leaves and short spikes of red-orange flowers. It is found in many dry habitats throughout the state.

*Triphysaria* (cream sacs and others) are annuals with deeply dissected leaves, mostly uncolored bracts, and a lower lip of 3 colorful, inflated petals.

*T. eriantha* (johnny-tuck) has red-tinted bracts and inflated yellow petals. It is common in grasslands.

*Orthocarpus* includes a few annuals with flowers similar to *Triphysaria*. (Formerly, most of the annuals in the last two genera were included in this genus.)

*Pedicularis* (louseworts) are perennials with often dissected, fernlike leaves and spikes of red, pink, or purple flowers with an upper lip that is snoutlike, hatchet-shaped, or curled like an elephant's trunk.

*P. densiflora* (Indian warrior) has deep red flowers and lives in foothill forests.

*P. semibarbata* (mountain lousewort) has yellow-orange flowers partly hidden by leafy bracts and lives in mountain forests.

*P. groenlandica* (elephant snouts) has pink-purple flowers with a galea resembling an elephant's trunk; it lives in wet mountain meadows.

*Cordylanthus* (pelican beaks) are annuals with leaves divided into threadlike segments and inflated white, cream-colored, or pinkish flowers whose lower lip resembles a pelican's pouched beak.

GENERA WITH A MORE CONVENTIONAL, 2-LIPPED DESIGN *Digitalis purpurea* (foxglove) is an introduced biennial with large, wrinkled, elliptical leaves and spikes of tubular pink, purple, or white flowers.

*Scrophularia* (figworts; bee plants) are perennials with pairs of triangular leaves and panicles of tiny maroon to almost black flowers with an overarching upper lip.

*Mimulus* (monkeyflowers; see fig. c) are small shrubs, perennials, and annuals with a sensitive 2-lobed stigma that closes when touched, and 4 stamens of 2 lengths.

*Penstemon* (penstemons; beard-tongues; see fig. d) are woody perennials with blue, purple, pink, red, or white flowers, 4 fertile stamens, and a 5th sterile stamen (no anther).

*Keckiella* (shrub penstemons) are similar but are full-fledged shrubs.

*Collinsia* (Chinese houses; tincture plant, blue-eyed Mary) are annuals with clusters or whorls of blue or purple flowers (often 2-toned), and what appear to be 4 petals; the 5th petal is hidden between the 2 apparent lower petals.

*C. heterophylla* (Chinese houses) has bicolored blue and white flowers and lives in foothill woodlands.

*C. tinctoria* (tincture plant) has pale purple to whitish flowers with intricate purple markings, foliage that stains skin, and lives in Coast Range and Sierra foothill woodlands.

*C. torreyi* (blue-eyed Mary) has tiny, bicolored blue and white flowers and lives in open mountain forests.

GENERA WHOSE 2 LIPS ARE FULLY CLOSED (NO OBVIOUS ENTRANCE TO THE FLOWER TUBE) *Antirrhinum* (snapdragons) are annuals and short-lived perennials with blue, purple, pink, or white flowers. They often appear in abundance after fire.

*Galvezia speciosa* (island snapdragon) is a shrub with whorled leaves and bright red flowers.

*Linaria canadensis* (native toadflax) is an annual with blue flowers and a tapered nectar spur.

#### a. *Veronica anagallis-aquatica* (speedwell or brooklime)

#### b. *Castilleja foliolosa* (woolly paintbrush)

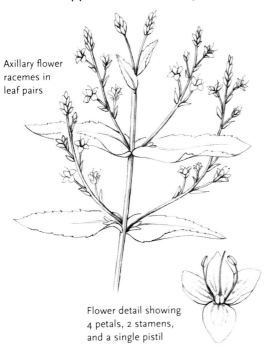

Axillary flower racemes in leaf pairs

Flower detail showing 4 petals, 2 stamens, and a single pistil

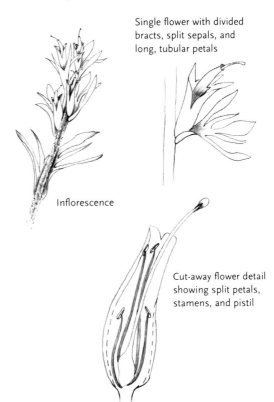

Single flower with divided bracts, split sepals, and long, tubular petals

Inflorescence

Cut-away flower detail showing split petals, stamens, and pistil

#### c. *Mimulus guttatus* (golden or seep monkeyflower)

#### d. *Penstemon heterophyllus* (blue foothill penstemon)

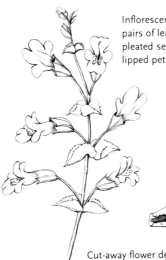

Inflorescence showing pairs of leafy bracts, pleated sepals, and 2-lipped petals

Cut-away flower detail showing spotted lower lip, 2 pairs of stamens, and a single style

Flower detail showing 2-lipped corolla, 4 fertile stamens, and single style

Seed pod

SCROPHULARIACEAE (FIGWORT OR SNAPDRAGON FAMILY)

# SOLANACEAE (Potato or Nightshade Family)

**RECOGNITION AT A GLANCE** Plants with often foul-smelling leaves; saucer-, trumpet-shaped, or tubular flowers with pleated petals; and often a fleshy berry in fruit.

**VEGETATIVE FEATURES** Herbaceous annuals, perennials, and small shrubs with simple, alternate, often foul-smelling leaves.

**FLOWERS** Usually showy, saucer- or funnel-shaped or tubular, and in varied arrangements.

**FLOWER PARTS** 5 partly fused sepals, 5 mostly fused petals pleated in bud, 5 stamens joined to the petals and sometimes stuck together, and a single pistil with a superior, (usually) 2-chambered ovary.

**FRUITS** Capsules and berries with many seeds.

**RELATED OR SIMILAR-LOOKING FAMILIES** The flower design of many nightshades resembles morning glories (Convolvulaceae), including the pleated petals, but differences include the milky juice, vininess, and few seeds per capsule typical of the morning glory family. A few members of the figwort family (Scrophulariaceae), such as mulleins (*Verbascum* spp.), have nearly regular flowers that also suggest this family, but scrophs do not have pleated petals and never have fleshy berries.

**STATISTICS** About 3,000 species of worldwide distribution, most diverse in the tropical Americas. The family is famous for its toxins, including several alkaloids such as scopolamine, hyoscyamine, nicotine, and belladonna, which in tiny doses are used medicinally. The family also has several major food plants including potatoes, tomatoes, tomatillos, chile and bell peppers, and eggplant. Several ornamentals, particularly shrubs, include cestrums, salpiglossis, angel's trumpets (*Brugmansia* spp.), yesterday-today-and-tomorrow (*Brunfelsia*), *Lycianthes*, petunias, and many more.

## CALIFORNIA GENERA AND SPECIES

The region has nine native or partly native genera and four nonnative genera.

### GENERA WITH DRY CAPSULES

*Datura* (jimson weeds, angel's trumpets, thornapples) are mostly perennials with large, fragrant, funnel-shaped, white to pale purple flowers that open at night.

*D. wrightii* (see fig. a) is widespread in dry areas and has flowers over 6 inches long.

*Nicotiana* (wild tobaccos) are annuals or perennials with foul-smelling, often sticky leaves; and slender, tubular to narrowly trumpet-shaped, whitish flowers.

*N. glauca* (tree tobacco) is an introduced South American species that is a tall shrub with blue-green leaves and long, tubular, yellow flowers.

*Petunia parviflora* (wild petunia) has sticky leaves on low, sprawling plants with single, purple, salverform flowers.

### GENERA WITH FLESHY BERRIES

*Lycium* (desert-thorns, frutillas) are dryland shrubs with ephemeral leaves and clusters of white or pale purple, trumpet-shaped flowers.

*Solanum* (nightshades) are annual, or perennial herbs, or small shrubs with white, purple, blue, or yellow flowers with wheel-shaped or recurved petals.

*S. umbelliferum* (see fig. b) and *S. xantii* (blue witch) are small, similar shrubs with blue-purple flowers and are widespread in California's foothills.

*S. nigrum* (black nightshade) is an introduced annual with small white flowers and swept-back petals.

*Chamaesaracha nana* is a sprawling perennial with white flowers similar to solanums but with short rather than long anthers.

*Physalis* (ground-cherries; tomatillos) are herbaceous plants with yellow flowers similar to solanums but with fruits enclosed in inflated, lanternlike sepals.

### a. *Datura wrightii* (angel's trumpet or thorn-apple)

Leaves, flower, and bud

Ripe capsule

### b. *Solanum umbelliferum* (blue witch)

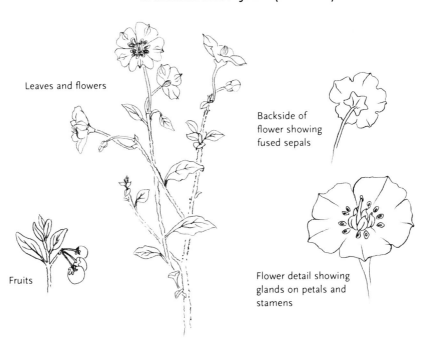

Leaves and flowers

Fruits

Backside of flower showing fused sepals

Flower detail showing glands on petals and stamens

SOLANACEAE (POTATO OR NIGHTSHADE FAMILY)

# TAXACEAE (Yew Family)

**RECOGNITION AT A GLANCE** Shrubs and trees with needlelike leaves and seed cones consisting of a single large seed surrounded by a fleshy wrapping.

**VEGETATIVE FEATURES** Dioecious, highly branched shrubs and slow-growing trees with spirally arranged, needlelike leaves (the leaves are often borne in two ranks or rows). The leaves are either unscented or have an unusual fragrance. The bark in mature specimens is scaly and brown to reddish brown.

**FLOWERS** Missing; these are cone-bearing plants.

**FLOWER PARTS** Tiny male pollen cones are catkinlike, cream colored, and borne in leaf axils. Female cones consist of a single large, poisonous seed surrounded by a cuplike red aril or completely enclosed in a prunelike purple aril.

**FRUITS** Mature seed cones resemble fruits because of their fleshy arils that attract birds.

**RELATED OR SIMILAR-LOOKING FAMILIES** The yew family is the only California conifer family with unusual, fleshy seed cones. The podocarp family (Podocarpaceae), largely confined to the southern hemisphere, also features fleshy seed cones but these are not of the same construction.

**STATISTICS** A tiny family of 16 species mostly confined to the northern hemisphere. Many yews (*Taxus* spp.) are commonly used as shrubs, hedges, and foundation plants in gardens, and are famed for their hard wood used in making bows. An anticancer agent, taxol, is now extracted from the needles of the western yew.

## CALIFORNIA GENERA AND SPECIES

The region has two native species and genera.

*Taxus brevifolia* (western or Pacific yew) has short, odorless, dark green needles that lie flat in two rows on the twigs. The needles are reminiscent of coast redwoods. Yew trees grow as large, horizontally branched shrubs or small trees; lack the fibrous, red-brown bark of redwoods; and the needles have a tiny ridge running down the upper center. The seed cones are unmistakable: a single seed sits in a fleshy, bright red cup. Western yew is uncommon and restricted to moist canyons in mid-elevation forests. Look for it in Calaveras Big Trees State Park, in protected canyons in the northern Sierra, and across the Klamath Mountains.

*Torreya californica* (California nutmeg or stinking-yew; see fig.) is a taller tree with gracefully drooping branches. Its long, glossy needles are arrayed in two rows and have wickedly sharp tips. The crushed needle has a strong and—to some—unpleasant odor. The purplish seed cones are the size of small prunes and the fleshy aril completely covers the seed. The common name alludes to the nutmeg-shaped seed. (The true nutmeg is a broadleaf, tropical tree in the Myristicaceae.) California-nutmeg grows in localized bands in conifer and mixed-evergreen forests in the central and north Coast Ranges and Sierra foothills.

## *Torreya californica* (California nutmeg)

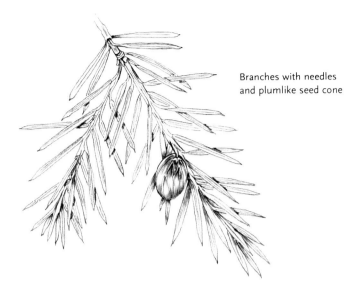

Branches with needles and plumlike seed cone

Sharp needles and pollen cones

# TAXODIACEAE (Redwood or Bald-Cypress Family)

RECOGNITION AT A GLANCE Conifer trees with awl- or needlelike leaves shed attached to a twig, and small, woody seed cones with spirally arranged, often diamond-shaped scales.

VEGETATIVE FEATURES Monoecious deciduous or evergreen trees with (usually) alternate, awl- or needlelike leaves.

FLOWERS Missing; these trees produce pollen and seed cones.

CONES Tiny pollen cones are in varied arrangements; small, woody seed cones have spirally arranged, often diamond-shaped scales and many small, winged, wind-dispersed seeds.

RELATED OR SIMILAR-LOOKING FAMILIES The redwood family might be confused with the pine family (Pinaceae) because of the needlelike leaves, but the leaves are shed attached to a twig, while needles in the pine family are shed singly, or, in the cases of pines (*Pinus* spp.), attached to minute spur shoots. Also, the seed cones of many Pinaceae are larger. The redwood family is closely related to the cypress family (Cupressaceae) and some botanists consider the two families as one. But members of the cypress family typically have scalelike leaves (mature foliage), and both the leaves and seed cone scales are in pairs or whorls.

STATISTICS Ten relict genera and species with an odd distribution: bald-cypresses (*Taxodium* spp.) in the southeastern United States and Mexico; two genera occur in Japan; several in China; and two species live in Tasmania. The coast redwood is of great economic importance for its beautiful and rot-resistant wood, and has unfortunately been overlogged in most parts of its range. Thankfully, the wood of our giant sequoia is considered too brittle to be of practical use. Several species are grown as ornamental trees, including the Japanese-cedar (*Cryptomeria japonica*), the bald-cypresses (*Taxodium* spp.), the dawn redwood (*Metasequoia glyptostroboides*), and the umbrella-pine (*Sciadopitys verticillata*).

CALIFORNIA GENERA AND SPECIES
The region has two native genera and species.

*Sequoia sempervirens* (coast redwood; see fig.) is the world's tallest tree, with specimens measuring over 370 feet high. It features dark red-brown bark, basal stump sprouts, flat needles in 2 rows (lower branches), and seed cones seldom over 2 inches long. Coast redwood dominates large tracts of foggy hills and flood plains, mostly in northern California.

*Sequoiadendron giganteum* (giant sequoia, Sierra redwood, bigtree) is the world's bulkiest tree, with circumferences up to 40 feet and heights over 250 feet. It features spongy, cinnamon-colored bark; sharp, awl-shaped leaves; and seed cones to 4 inches long. Giant sequoia lives in mixed groves with other members of the yellow pine forest from the central Sierra south. Dramatic groves can be seen in Yosemite, Kings Canyon, and Sequoia national parks.

### *Sequoia sempervirens* (coast redwood)

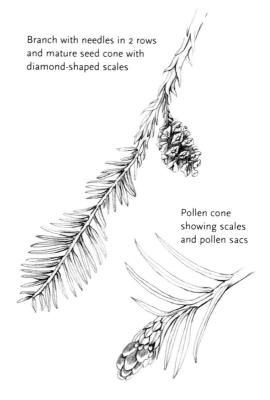

Branch with needles in 2 rows and mature seed cone with diamond-shaped scales

Pollen cone showing scales and pollen sacs

# THEMIDACEAE (Brodiaea Family)

RECOGNITION AT A GLANCE Corm-bearing plants with bracted umbels of flowers and tepals (usually) joined to form a petal tube.

VEGETATIVE FEATURES Fibrous-coated corms send up a few long, strap-shaped, unscented leaves that often wither by flowering time.

FLOWERS Small to medium-sized; trumpet-, star-, or bell-shaped; and arranged in bracted umbels.

FLOWER PARTS 2 rows of 6 tepals usually joined partway to form a petal tube; 3 or 6 stamens (sterile stamens or staminodes often present); and a single pistil with a superior, 3-chambered ovary.

FRUITS 3-chambered capsules.

RELATED OR SIMILAR-LOOKING FAMILIES This family has been recently separated from others with superficially similar appearance. Previously its members were usually placed in the amaryllis family (Amaryllidaceae), lily family (Liliaceae), or onion family (Alliaceae). The onion family is easily separated by the strong onion odor in the leaves and the separate tepals (joined in a very few species).

STATISTICS Around 50 species from western North America, Mexico, and Central America. A few are cultivated as ornamentals in western gardens. Corms of several were dug and eaten by the California Indians.

## CALIFORNIA GENERA AND SPECIES

The region has six native genera and between 35 and 40 species.

### GENERA WITH A DEFINITE PETAL TUBE

*Brodiaea* (brodiaeas) have waxy tepals and 3 fertile stamens alternating with 3 petal-like, sterile stamens. Most of the species live in open habitats, bloom late, and feature blue, purple, or pink flowers. Several species are highly restricted in distribution and favor unusual, ancient soils.

*B. elegans* (elegant brodiaea) is widespread and features blue to blue-purple flowers.

*B. californica* (California brodiaea) grows in the northern Sierra foothills and inner north Coast Ranges. It has flowering stalks to 2 feet high with narrow, pale pink-purple flowers.

*B. terrestris* (ground or dwarf brodiaea) is scattered along the coast and inland in valley bottoms of the Coast Range foothills. Its flowering stalks appear to come directly out of the ground and the flowers are pale blue.

Most of the true brodiaeas are told apart by details of the flower tube and shape, placement, and color of the sterile staminodes. A new species has been recently described from the region of the Santa Rosa Basalt Area of the Santa Ana Mountains in southern California.

*Dichelostemma* (blue dicks, firecracker flower, and others) have unwaxy tepals, bent or twisted flowering stalks, and either 6 fertile stamens (with 2 different sizes of anthers) or 3 fertile stamens backed by petal-like appendages. The five species (and one hybrid) usually live in partly shaded habitats and bloom from early spring to early summer.

*D. capitatum* (blue dicks; see fig.) is perhaps the most widespread of all native species and blooms in early to mid-spring. It features head-like clusters of blue, purple, or pinkish flowers and 6 stamens of two alternating sizes.

*D. volubile* (twining brodiaea) is a strange plant with stems that twine around other vegetation for support. It lives in the Sierra and north Coast Range foothills and features pink flowers.

*D. ida-maia* (firecracker brodiaea) has unusual hanging, tubular, crimson-red flowers with recurved green tepals. This hummingbird-pollinated flower favors edges of redwood and other conifer forests in northern California's Coast Ranges and Klamath Mountains.

*Triteleia* (Ithuriel's spear, golden brodiaea, and others) have unwaxy tepals and 6 fertile stamens with or without small appendages. Triteleias live in a variety of habitats and feature yellow, white, purple, pink, or blue flowers that bloom from midspring into early summer.

*T. laxa* (Ithuriel's spear) has showy blue or purple flowers and is widespread in the foothills.

*T. hyacinthina* (white brodiaea) has shallow white to purple-flushed flowers and is common in temporarily wet grasslands in the foothills.

*T. ixioides* (pretty face or golden brodiaea) has star-shaped, pale to bright yellow flowers and is found in various forms from the central coast through the Sierra foothills and on into meadows in the high mountains.

*Androstephium breviflorum* has 6 fertile stamens whose filaments are fused to form a crown. It is very rare in our eastern high deserts.

GENERA WITHOUT AN APPARENT PETAL TUBE

*Muilla* (four species) have tiny, star-shaped, white or pale yellow flowers.

*M. maritima* is widespread in the grassy foothills but is seldom noticed due to its small flowers.

*Bloomeria* (golden stars, two species) have bright yellow flowers and pairs of appendages at the base of the stamens. They live in woodlands and chaparral in central and southern California.

*B. crocea* is common and widespread in open situations from Monterey County south through the rest of central and southern California's mountains.

### *Dichelostemma capitatum* (blue dicks brodiaea)

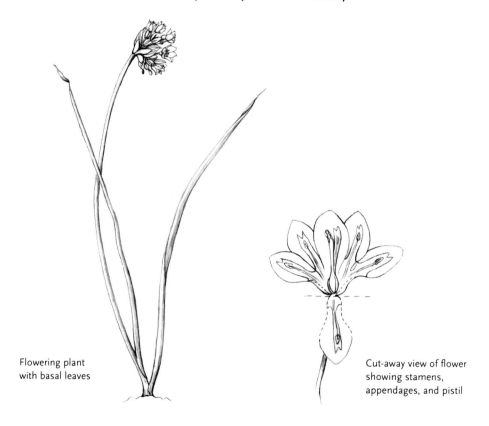

Flowering plant with basal leaves

Cut-away view of flower showing stamens, appendages, and pistil

# VERBENACEAE (Verbena Family)

RECOGNITION AT A GLANCE Plants with square stems and simple, opposite leaves that are sometimes fragrant but not in the manner of mints. Flowers are slightly irregular and arranged in heads or spikes.

VEGETATIVE FEATURES Ground covers, herbaceous perennials, and small shrubs with square stems and opposite, simple (sometimes deeply lobed) leaves that are slightly scented but without the strong sagey or minty fragrance typical of the Lamiaceae.

FLOWERS Small, slightly irregular, purple or pinkish, and arranged in headlike clusters or spikes.

FLOWER PARTS 5 partly fused sepals, 5 somewhat unequal petals joined to form a tube, 4 stamens, and a single pistil with a 2- to 4-lobed, superior ovary.

FRUITS Drupes, capsules, or nutlets.

RELATED OR SIMILAR-LOOKING FAMILIES The verbena family is closely related to and sometimes confused with the mint family (Lamiaceae). A combination of characters usually separates the two: mint leaves are more highly aromatic; mint flowers are usually strongly irregular and two-lipped; mint fruits always develop into 4 one-seeded nutlets. Verbena fruits vary considerably. By and large, the verbena family is tropical, and California has few species, whereas the mints are more diverse in temperate regions.

STATISTICS 1,900 species found mostly in the tropics; a few are cultivated in our gardens, including lemon verbena (*Aloysia citriodora*), lantanas (*Lantana* spp.), and verbenas (*Verbena* spp.). The family is also noteworthy for the hardwood teak (*Tectona grandis*).

## CALIFORNIA GENERA AND SPECIES

The region has three mostly native genera with 13 species, and two nonnative genera with three species.

### NATIVE GENERA

*Verbena* (vervain, verbena, seven native species and three introduced species) has nearly regular petals and five sepals.

*V. lasiostachys* (common vervain) grows along watercourses and in dry stream beds and has narrow spikes of pale purple flowers.

*V. hastata* (delta verbena; see fig.) occurs in marshy places in the Sacramento Delta and has tall stems with slender spikes of purple flowers.

*V. gooddingii* (desert verbena) lives on sandy flats in the high desert. Its low mounds have headlike clusters of showy lavender blossoms.

*Phyla* (2 native species) has 2 or 4 sepals and irregular, 2-lipped petals.

*P. nodiflora* is widespread in moist places and forms sprawling mats with whitish flowers.

### NONNATIVE GENERA

*Avicennia marina* var. *resinifera* (black mangrove) has moved into coastal salt waters near San Diego. It comes from New Zealand and is among a group of tropical trees called *mangroves,* adapted to saltwater sloughs.

*Lantana camara* (lantana) in coastal southern California. Lantana is a common South American shrub with red to yellow flowers attractive to butterflies.

# VERBENACEAE (Verbena Family)

## *Verbena hastata* (delta verbena)

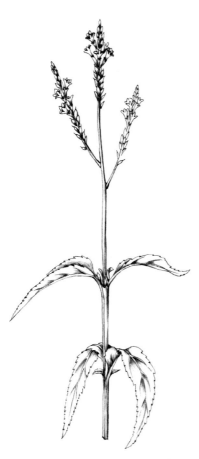

Upper branch showing square stems, opposite leaves, and slender flower spikes

Fruiting spike showing sepals around fruits

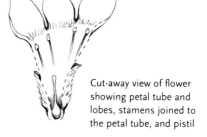

Cut-away view of flower showing petal tube and lobes, stamens joined to the petal tube, and pistil

# VIOLACEAE (Violet Family)

RECOGNITION AT A GLANCE Perennials with basal leaves and solitary, irregular, yellow, blue, or white, violetlike flowers with a nectar sac or spur.

VEGETATIVE FEATURES Herbaceous perennials with basal, heart-shaped to deeply lobed leaves and stipules.

FLOWERS Showy, irregular, spurred; yellow, white, or blue; and usually solitary above the leaves.

FLOWER PARTS 5 separate sepals; 5 separate, irregular petals—the lower middle petal extends back into a nectar sac or spur—5 fused stamens at the entrance to the spur; and a single pistil with a superior, 3-sided, 1-chambered ovary.

FRUITS Explosive capsules with several seeds.

RELATED OR SIMILAR-LOOKING FAMILIES Although violet flowers are irregular, the petals are not joined to create a 2-lipped design, nor are they arranged in a papillinaceous pattern as in the pea family (Fabaceae). The middle lower petal usually carries stripes or lines that serve as nectar guides. Many of the tropical genera of the family look utterly different, and are woody shrubs and small trees with regular flowers.

STATISTICS 600 species distributed throughout the world, including the tropics, where shrub- and treelike forms occur. The family has long been valued for ornamentals in the genus *Viola*, including English violets, violas, and pansies. The flowers of some violets are candied or used fresh in salads.

## CALIFORNIA GENERA AND SPECIES

The region has one mostly native genus and 24 species.

*Viola* (wild violets, pansies) are mostly native perennials, with several species distributed in a wide variety of habitats. Species may be sorted according to flower color or leaf design.

SPECIES WITH NONYELLOW FLOWERS *Viola ocellata* (western heartsease) has white petals trimmed with purple stripes and 2 yellow eye spots.

*V. macloskeyi* (white meadow violet) has white flowers with twisted petals.

*V. adunca* (dog violet) has pale to deep blue flowers on creeping plants.

SPECIES WITH YELLOW FLOWERS AND LOBED LEAVES *Viola douglasii* (Douglas's violet) has pansylike flowers and lives in foothill woodlands and grasslands.

*V. sheltonii* (Shelton's violet) has small flowers and grows in montane forests on scree.

*V. lobata* (lobed violet) has large flowers backed brown, and leaves with only a few lobes; it lives in mixed-evergreen and yellow pine forests.

SPECIES WITH YELLOW FLOWERS AND UNLOBED LEAVES *Viola purpurea* (pine violet) has narrow, grooved leaves and lives in many wooded habitats.

*V. glabella* (smooth yellow violet) has broad leaves and lives in moist conifer forests.

*V. sempervirens* (redwood violet; see fig.) has creeping stems and round leaves and lives in coastal conifer forests.

*V. pedunculata* (wild pansy) has round leaves and large, butterscotch-yellow flowers. It lives in foothill grasslands and woodlands.

## *Viola sempervirens* (redwood violet)

Habit showing roots, nearly round leaves, and solitary flower on leafless stalk

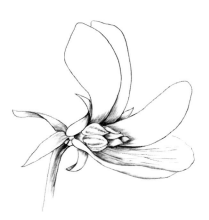

Cut-away flower detail showing narrow sepals, irregular petals, and stamens forming a cone around pistil

# GLOSSARY

ACCESSORY FRUIT — A fruit made by a fleshy receptacle engulfing the pistils. Example: strawberry fruit.

Achene

ACHENE — A single-seeded fruit. The fruit is shed as though it were a seed. A good example is an unshelled sunflower "seed."

ACORN — A nut that sits in a scaly or warty cup. Example: the fruits of oak trees.

AGGREGATE FRUIT — A fruit composed of several small, fleshy drupes. Examples: blackberry and raspberry fruits.

Alternate (leaves)

ALTERNATE — Refers to leaves attached at different levels on stems.

ANNUAL — A plant that lives for one year or less. Examples: baby-blue-eyes and clarkias. Many wildflowers are annuals.

Anther (stamen)

ANTHER — The pollen sac at the end of a stamen.

APPENDAGE — A general term that refers to extra flaps of tissue attached to petals and other structures.

Appendage

AQUATIC — Describing plants that live in or under water.

ARIL — A fleshy, brightly colored wrapping or attachment to a seed that attracts birds. Example: the red cup around a yew seed.

AURICLE — Down-turned green flaps between sepals, such as those on nemophila and fiesta flowers.

| | | |
|---|---|---|
| | AWN | A spikelike extension on a grass grain or the achene of a composite. Awns aid in seed dispersal. |
| | AXIL (ADJ. AXILLARY) | The upper angle between a leaf and the stem it is attached to. |
| | BANNER | The upper back petal of a typical sweetpea type (papilionaceous) flower (illustration under *papilionaceous*). |
|  Basal (leaves) | BASAL | Describing leaves arranged at the base of a plant. |
| | BASIFIXED | Describing an anther that is firmly attached at the base to its filament. |
|  Berry | BERRY | A fleshy fruit containing several seeds. Examples: tomatoes, bananas, and eggplants. |
| | BIENNIAL | A plant that lives for two years. The first year, it makes a leaf rosette, the second year, it flowers. Example: mulleins (*Verbascum* spp.). |
| 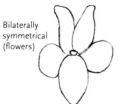 Bilaterally symmetrical (flowers) | BILATERALLY SYMMETRICAL | Describing flowers whose petals are not all the same size or shape. Such flowers can only be cut into two in one direction to produce matching halves. |
| | BISEXUAL | Describing flowers that have both stamens and pistil(s). |
| | BLADE | The flattened part of a leaf. |
|  Bract | BRACT | Any modified leaf associated with a flower. Bracts may be green like small leaves, or they may be colorful and replace petals (as with the calla lily and poinsettia). |
| | BUD | The resting stage or immature, undeveloped structure that later opens to a flower or grows into a shoot. Buds can be *terminal* at the tip of a stem or *lateral* in the axils of leaves. |
| 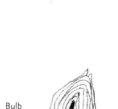 Bulb | BULB | Ball- or globe-shaped underground organ that stores food and water during dormancy. Consists of fleshy leaf bases around a bud. Example: onion. |
| | CALYX | The collective term for the sepals of a flower. |
| Capsule | CAPSULE | A dry seed pod that splits into two or more sections to shed its seeds. Examples: the seed pods of poppies and monkeyflowers. |

| | |
|---|---|
| CARYOPSIS | The grain of a grass, which is the ovary fused to the seed inside. Examples: corn kernels and wheat berries. |
| CATKIN | A long, narrow chain of petalless greenish or brownish flowers that are wind-pollinated. Examples: the male flowers of oaks and hazelnuts. |
| CAUDEX | Substantial basal stem. |
| CHAPARRAL | Drought-adapted plant community of mostly evergreen shrubs that live on hot, dry, rocky slopes. |
| CILIA | Whiskerlike hairs on leaves, petals, and other structures. |
| CLAW | The narrowed base of a petal or other flower part. Example: petals in the mustard family. |
| COLUMN | The complex structure in orchid flowers that consists of a stamen(s) fused to the style and stigma. |

Column (orchid flowers)

| | |
|---|---|
| COMPOUND | Describing leaves composed of several distinct leaflets. A bud sits at the base of the leaf, but not at the base of the leaflets. Compound leaves may be pinnately compound, palmately compound, or trifoliate. |
| CONIFER | A shrub or tree that bears its seeds in cones and has needlelike or scalelike leaves. Examples: pines, firs, redwoods, and junipers. |
| CORM | An underground organ similar to a bulb but solid inside (modified starch-storing stem rather than leaf bases). Example: Gladiolus "bulb." |

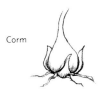
Corm

| | |
|---|---|
| COROLLA | The collective term for petals of a flower. |
| CORONA | A raised crownlike rim between the petal lobes and petal tube. Example: forget-me-not flowers. |
| COTYLEDON | The leaves of a seedling. The number of cotyledons determines whether the plant is a monocot or dicot. |
| CYME | A complex arrangement of flowers in flat-topped clusters. |

Cyme

| | |
|---|---|
| DECIDUOUS | Describing shrubs and trees that lose all their leaves at one time of the year. |

GLOSSARY 181

DICHOTOMOUS | Describing branches and veins that are equally forked.

DICOTS | The largest group of flowering plants, usually identified by branched or netlike leaf veins and sepals and petals in fours or fives.

DIOECIOUS | Describing male and female flowers borne on separate plants. Examples: cottonwood trees and silk-tassel bushes.

Disc flower (Asteraceae)

DISC FLOWER | Describes the tiny, star-shaped flowers in the middle of a composite or daisy head. Each flower has five symmetrical petals, five fused stamens, and two styles.

DISSECTED | Refers to leaves or other structures that are deeply divided into many pieces. Examples: bleeding heart and California poppy leaves.

DISTURBED | Describing habitats that are altered from their natural state, such as roadsides, grazed pastures, vineyards, and orchards.

DOMINANT | Refers to the most conspicuous plants in a plant community, such as redwoods in a redwood forest. Also, in life cycles, the obvious part of the cycle, such as the spore-bearing fern plant, or the leafy cushion of a moss.

DRUPE | A fleshy fruit with a single large pit, as in peaches and avocadoes.

ENDEMIC | Describing plants that are naturally restricted to one particular place.

ENDOSPERM | The stored food inside a seed.

ENTIRE | Describing smooth, even leaf margins without teeth or lobes.

Entire (leaves)

EPIDERMIS | The outer skin of stems, roots, and leaves.

EQUITANT | Refers to flattened sprays of leaves that overlap at the base. Example: iris leaves.

ESCAPED | Describing plants that grow on their own away from gardens.

EVERGREEN | Describing shrubs and trees that keep some leaves the entire year.

EXSERTED | Describing stamens and styles that extend beyond the petals of a flower, as on fuchsia flowers.

| | | |
|---|---|---|
| | FAMILY | A group of related genera that share certain features. Families may be small or very large. Family names end in *-aceae*. |
| | FASCICLE | A cluster of leaves or other structures. Example: the leaves of chamise. |
| | FERTILE | Refers to flowers or parts of flowers that have the ability to produce pollen or seeds. |
| Filament (stamens) | FILAMENT | The stalk of a stamen. |
| | FLORET | The tiny flowers of grasses (Poaceae) and composites (Asteraceae). |
| Follicle | FOLLICLE | A seed pod with a single chamber that opens by one lengthwise slit. Examples: columbine and milkweed seed pods. |
| | FRUIT | General term for the ripe ovary of flowering plants. Fruits may split open to release seeds or be fleshy or woody. |
| | GENUS (PL. GENERA) | A group of related species that share many traits. Often used in common language to denote important groups like oaks, roses, and pines. |
| | GLAUCOUS | Bluish green. |
| | GLUME | The bract or bracts at the base of grass spikelets. |
| | GYNOECIUM | The collective term for the pistils of a flower. Most flowers have a single, compound pistil, but primitive flowers have a gynoecium of several separate pistils. |

| | | |
|---|---|---|
| Gynostegium (milkweeds) | GYNOSTEGIUM | The complex reproductive structures of milkweeds in which the stigmas, styles, and stamens are fused together. |
| | HABIT | The overall form or shape of a plant, such as tree, shrub, ground cover, etc. |
| | HABITAT | The environment in which plants live. Do not confuse with habit. |
| Head | HEAD | A tight cluster of flowers at the end of a single stem. Example: daisy "flowers." |
| | HERBACEOUS (HERB) | Describing plants that are not woody. |
| | HOOD | The nectar cups of milkweed flowers, which are modified outgrowths of the stamens. |

GLOSSARY 183

| | |
|---|---|
| HYBRID | A cross between two genetically different individuals. Hybrids may occur between different varieties, subspecies, and sometimes species. |
| HYPANTHIUM | Flower tube that bears sepals, petals, and stamens at the top. Examples: the flower tubes of evening-primrose and currant flowers. |
| INFERIOR | Describing ovaries that are positioned below the other flower parts, as in fuchsias and currants. |
| INFLORESCENCE | The arrangements of flowers. |
| INTRODUCED | Refers to plants that have been deliberately or accidentally brought to a new home, where they grow on their own. |
| INVOLUCRE | A cup-shaped or bowl-shaped row of bracts around a group of flowers. Example: the vase- or turban-shaped structures that hold the flowers of wild buckwheats. |
| JOINTED | Describes stems and other parts of plants that are pinched in between sections or break apart. |
| KEEL | The boat-shaped middle petals of a sweetpealike or papilinaceous flower. Also the V-shaped midrib of a leaf. |
| LEADER | The main growing tip of a tree. |
| LEAFLET | The individual parts of compound leaves. Often leaflets are mistaken for simple leaves but they never have a bud or stipules at their base. |
| LEGUME | A member of the pea family (Fabaceae); also their fruit. A legume is a single-chambered seed pod that opens by two long slits (as with a pea pod). Compare to follicle. |
| LEMMA | The upper or back bract of a grass floret. |
| LIGULE | A tongue-shaped structure. Used to refer to the petals of the ray flowers of composites, and a small appendage at the base of grass leaves. |
| LINEAR | Having a long, narrow shape, with parallel sides. |
| LIP | The single enlarged lower petal of orchid flowers or one of two sets of petals at the entrance to a flower tube in two-lipped flowers, as in flowers of the snapdragon family (Scrophulariaceae). |

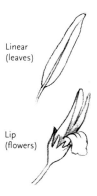

Lip (flowers)

Lobed

| | | |
|---|---|---|
| | LOBED | Describing leaves or petals with deep, smooth indentations. |
| | MONOCOTS | The other major group of flowering plants (see *dicots*), characterized by leaves with parallel veins and flowers with sepals and petals in threes. |
| | MONOECIOUS | Used to describe male and female flowers produced on the same plant, as on oaks and alders. |
| | MUCRO | A tiny spinelike tip of a leaf or petal. |
| | MYCOPARASITE | A plant that parasitizes fungi. Examples: coralroot orchids and snowplant. |
| | NATIVES | Plants that are naturally occurring and were not introduced by humans. |
| | NATURALIZED | Said of nonnative plants that appear to be native because they are growing on their own without help. |
| | NECTAR | Solution of sugar water that flowers produce to attract and reward pollinators. |

Node

| | | |
|---|---|---|
| | NODE | The place where the leaf joins the stem. |
| | NUMEROUS | Used in botany to describe the number exceeding 12 of some flower part, such as the stamens in buttercup flowers. |
| | NUT | Large, single-seeded fruit with a hard shell. Examples: hazelnut, oak acorn. |

Nutlet

| | | |
|---|---|---|
| | NUTLET | A small, single-seeded fruit that does not open. Similar to an achene, but with a thick and hard ovary wall. Examples: fruits of the mint family (Lamiaceae) and borage family (Boraginaceae). |
| | OCHREA | Papery stipules typical of many species in the Polygonaceae. Example: leaves of docks and the edible buckwheat. |

Opposite (leaves)

| | | |
|---|---|---|
| | OPPOSITE | Describing leaves in pairs on stems. |
| | OVARY | The swollen bottom part of the flower's pistil that contains the seeds. |

Ovary

| | | |
|---|---|---|
| | OVATE | Describes plant parts that are round at the base, broadest just below the middle, and tapering to a point. |
| | PALEA | The lower or inner bract of a grass floret. |

Palmate

PALMATE  Having veins, leaf lobes, or leaflets arranged in a fanlike manner, like the fingers on a hand.

Panicle

PANICLE  Complex grouping of flowers consisting of a raceme of racemes. See raceme.

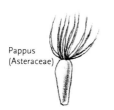

Papillionaceous (Fabaceae)

PAPILIONACEOUS  Describing a sweetpealike flower, in which the petals consist of an upper back *banner*, two side *wings*, and two fused middle boat-shaped petals (*keel*).

PAPPUS  The highly modified sepals of individual composite flowers. The pappus consists of hairs, bristles, or scales, such as the down on dandelion seeds.

Pappus (Asteraceae)

PARASITE  A plant or animal that gets its food from other living organisms. Examples: mistletoe on oak trees; broomrapes on grasses.

PEALIKE  See *sweetpealike* or *papilionaceous*.

PEDICEL  A small stalk that bears a flower.

PEDUNCLE  A larger stalk that bears (groups of) flowers.

PELTATE  Umbrella-shaped, referring to round leaves attached to a central petiole. Example: nasturtium leaves.

PEPO  Gourdlike fruit; a berry with a hard outer rind. Examples: squashes and melons.

PERENNIAL  A plant living three or more years.

PERFOLIATE  Referring to leaves joined around a stem. Example: the stem leaves of miner's lettuce.

PERIANTH  Collective term for the sepals and petals of a flower.

PERIGYNIUM  Saclike structure around the ovaries in sedge flowers and fruits.

Petal (flower)

PETAL  The second row of parts in a flower; usually brightly colored to attract pollinators.

PHOTOSYNTHESIS  The process in green plants that takes carbon dioxide from the air, combines it with water from the soil, and uses sunlight to produce sugar and oxygen.

PHYLLARY  The bracts around the flower heads of composites or daisies.

Pinnate

| | | |
|---|---|---|
| | PINNATE | Describes veins, leaf lobes, or leaflets arranged in a featherlike pattern (like the barbs along the shaft of a feather). |
| | PIONEERS | Plants that grow on newly disturbed sites. |
| | PISTIL | The center female part of the flower, which consists of the ovary, style, and stigma. |
| | POLLEN | The fine, powderlike, often yellow dust produced by stamens and used to pollinate flowers. |
| | POLLINATION | The process of transporting pollen from the anther to the stigma. This process is usually carried out by insects and birds. |
| | POLLINIUM | Masses of pollen lumped together in orchid and milkweed stamens. |
| | POME | A fruit with a papery ovary containing several seeds surrounded by a fleshy receptacle. Examples: apple and pear fruits. |
| | PRICKLE | Spinelike growth from the skin of stems not associated with leaf nodes. Examples: "spines" of roses and blackberries. |
| | PROSTRATE | Said of plants that creep along next to the soil. |
| | RACEME | A long stem bearing side branches, each with a flower. |
| | RADIAL | Describing flowers with petals all the same shape and size. |
| | RAY FLOWER | The large, showy outer flower of composites or daisies. The five petals are fused into a tongue-shaped structure that looks like a single petal. |
| | RECEPTACLE | The stem end to which all flower parts are attached. |
| | RHIZOME | Creeping underground stem thickened to store food. |
| | RIPARIAN | Referring to plants that grow on the flood plains of streams and rivers. |
| | ROOTSTOCK | Rootlike, underground stem that probes deeply into soil or under rocks. |
| | ROSETTE | A symmetrical set of basal leaves. |
| | SAMARA | Winged fruit. Examples: maple and ash fruits. |

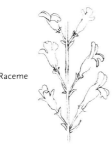
Raceme

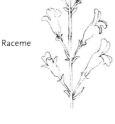
Radial (flowers)

Ray flower (Asteraceae)

Rhizome

Rootstock

Rosette

GLOSSARY 187

| | | |
|---|---|---|
| | SAPROPHYTE | A plant that feeds on dead organisms. Examples: fungi that live in the leaf litter layer of the forest. |
| | SCALES | Structures (such as the pappus of composite flowers) that are shaped like fish scales. |
| | SCAPE | Naked, flower-bearing stem that arises from basal leaves. |
| Schizocarp | SCHIZOCARP | Fruit that splits into sections, each with one or more seeds inside. Example: hollyhock and parsley fruits. |
| | SEPAL | The outermost whorl of the flower, covering and protecting the flower bud. |
| Sepal (flower) | SERPENTINE | Slick, ultramafic, bluish green rock common in California's foothills. Serpentine-derived soils have properties that most plants can not handle well. |
| | SERRATE | Having sawlike "teeth" (used to describe the margins of leaves and petals). |
| | SESSILE | A stalkless leaf or flower. |
| | SHRUB | A woody plant with multiple branches, not one main trunk. |
| Silicle (Brassicaceae) | SILICLE | Special fruit type in the Brassicaceae, where the ovary is as broad as long. Example: money plant seed pods. |
| Silique (Brassicaceae) | SILIQUE | Special fruit type in the Brassicaceae, where the ovary is much longer than broad. Examples: mustard and wallflower seed pods. |
| | SIMPLE | Describing a leaf that is not divided into leaflets. Simple leaves may be deeply lobed or slashed. |
| | SPECIES (SING. AND PL.) | The kinds of plants in a genus, as for example, blue oak, valley oak, and coast live oak in the genus *Quercus*. |
| Spike | SPIKE | Flowers borne directly along (up) a stem. |
| | SPINE | Sharply pointed structures that represent modified leaves. Spines always occur at the nodes of stems. Example: the spines on gooseberry stems. |
| Spur | SPUR | A narrow, pointed nectar sac, such as on nasturtium and columbine flowers. |

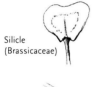

| | | |
|---|---|---|
| | SPUR SHOOTS | Very short side branches that bear flowers and fruits or clusters of leaves. Examples: apple and pear trees; needles in pines and cedars. |
| Stamen (flower) | STAMEN | The male part of the flower located just outside the pistil. Each stamen consists of a stalk (*filament*) and pollen sacs (*anthers*). |
| | STELLATE | Describing hairs on leaves and stems that look like starbursts (hand lens). |
| | STERILE | Refers to flowers and fruits that do not produce viable pollen or seeds. |
| Stigma (flower) | STIGMA | The top of the pistil, usually enlarged and often sticky or hairy to accept pollen. |
| | STIPE | Stalk that carries the ovary above the receptacle of the flower. |
| Stipules (leaves) | STIPULES | Pairs of (usually) small, leaflike structures at the base of leaves. Certain families typically have stipules; others do not. Examples: leaves in the pea and geranium families (Fabaceae and Geraniaceae). |
| | STOLON (IFEROUS) | Describing slender runners that carry new plantlets away from the parent plant, as with strawberry plants. |
| Style (flower)  | STYLE | A stalk at the top of the pistil's ovary that ends in a stigma. |
| | STYLOPODIUM | The enlarged style base in flowers of the parsley family (Apiaceae). |
| | SUBSPECIES | Races of plants within a species. |
| | SUCCULENT | Describing leaves or stems that store water, or plants with such leaves or stems. |
| | SUMMER-DORMANT | Refers to perennials that die back to their roots in the summer. |
| Superior (ovary) | SUPERIOR | Said of ovaries positioned above the attachment of the other flower parts. |
| | TAXONOMY | The study of classifying, naming, and identifying organisms. |
| Tendril  | TENDRIL | A coiled stalk that clasps onto other plants to allow vines to climb. Examples: grapes and wild cucumbers. |

| | | |
|---|---|---|
| TEPAL | | Sepals and petals that look alike, or a perianth whose parts all look the same. Example: lily flowers. |
| TERNATE | | Describing leaves divided into threes or multiples of threes. Ternate leaves are common in the Ranunculaceae. |
| THORN | | Spinelike, sharp-tipped side branch. Example: the "spines" of hawthorns. |
| TREE | | A woody plant with one or few main trunks and smaller side branches. |
| TRIBE | | A group of related genera that share certain features. Tribe is a category between family and genus. Examples: large families such as composites and grasses are divided into several tribes. |
| TRIFOLIATE | | Describing compound leaves divided into three leaflets. Examples: clover and strawberry leaves. |
| TRUNCATE | | Refers to a leaf or petal that is squared off. |
| TUBER | | An underground swollen stem that bears eyes. Example: potato. |
| TUBERCLE | | Wartlike protruberance on leaves, stems, and flowers. |

Two-lipped (flowers)

**TWO-LIPPED** Refers to bilaterally symmetrical flowers with petal lobes above and below the opening to the petal tube (throat). Examples: most flowers in the Scrophulariaceae and Lamiaceae.

Umbel

**UMBELS** Flowers borne at the ends of radiating spokelike stalks, like the ribs of an umbrella. Examples: members of the parsley family (Apiaceae).

**UNISEXUAL** Describing flowers with either stamens or pistils but not both. Example: squash flowers.

**UTRICLE** A small, inflated achene.

**VARIETY** Named variations within species. Similar to subspecies.

**VEGETATIVE** Refers to the roots, stems, and leaves but not the flowers or fruits.

Veins (leaves)

**VEINS** The vascular strands that carry water and food in leaves. Vein pattern is important in differentiating dicots from monocots.

| | |
|---|---|
| VERNAL POOLS | Depressions in valleys and foothills that fill with water in winter and spring and are dry in summer. Vernal pools are home to many highly specialized plants. |
| VERSATILE | Refers to an anther that swings or pivots on its filament. Example: stamens of lilies (*Lilium* spp.). |
| VOLATILE OILS | Small, fragrant terpenes that quickly evaporate. Such oils are common in flowers and leaves. |
| WHORLED | Used to describe three or more leaves attached at the same level of a stem. Example: *Lilium* leaves. |

Whorled (leaves)

# COMMON PLANT-NAMES INDEX

Bolded numbers indicate pages with figures.

acacia, 81
   catclaw, 81, **83**
acanth, 165
acanth family, 98
agave, 18
agave family, 18–19
alder, mountain, 44, **45**
   red, 44
   white, 44
alfalfa, 81, 82
alkali sacaton, 128
alkali weed, 63
all-spice, 111
almond, 152
alumroot, 162
   common, **164**
alyssum, sweet, 49
amaranth family, 61, 137
amaryllis family, 21, 173
amole, 104
anemone, 147
*Angelica*, 24
angel's trumpet, 168, **169**
anise, 24
antelope brush, 153
apples, 152
aralia family, 24
*Arnica*, 37
artichoke, globe, 34
   Jerusalem, 34
ash, flowering, 114
   Oregon, 114, **115**
   velvet, 114
asparagus, 103

asphodel, bog, 104
aster, 34
   golden, 35
   woolly, 35
aster tribe, 35
avens, 152
avocado, 101
azalea, 75
   western, 76, **77**

baby-blue-eyes, 92
bald-cypress, 172
bald-cypress family, 172
balsamroot, 36
bamboo, heavenly, 42
baneberry, 148
barberry, 42
   long-leaf, **43**
barley, 128
   wild, 129
barley tribe, 129
basil, 98
basketflower, golden, 48
bay, California, 101, **102**
bay-laurel, 101
   California, 101
bead-lily, 103
beard-tongue, 166
bear-grass, 18, 105
bedstraw, 156, **157**
beech family, 84–85
beechnut, 84
beeplant, 166

beet, 61
   sugar, 61
bellflower, 54
bellflower family, 54–55
bent-grass tribe, 128
berry-rue, 158
bignons, 165
bigtree, 172
bilberry, 75
bindweed, 63
birch, resin, 44
   water, 44
birch family, 44–45
birthwort family, 42, 101
biscuit-root, 25
bishop's cap, 162
bistort, 137
bitterroot, 141, **142**
bladderpod, 49, 56, **56**
blazing star, common, **107**
   satin, 107
blazing star family, 107
blechnum fern family, 145
bleeding heart, 122
   western, 122
bloodroots, 94
bluebells, mountain, 46, **47**
   Scottish, 54
blueberries, 75
bluecups, 54, **55**
blue dicks, 173, **174**
blue-eyed grass, 94, **95**
blue-eyed Mary, 46, 166
bluegrasses, 129
blue witch, 168, **169**
blow wives, 36
bluff lettuce, 66
borage, 46
borage family, 46, 47
bottlebrush, 111
bougainvillea, 113
bouncing bet, 60
box elder, 16, **17**
bracken family, 145
Brake fern family, 145, 146
breadroot, California, 82
breath-of-heaven, 158
bride's bouquet, 134
broccoli, 48
Brodiaea, California, 173
   dwarf, 173
   elgeant, 173
   firecracker, 173
   golden, 173, 174
   ground, 173

pretty face, 174
   twining, 173
   white, 174
Brodiaea family, 173–174
brome, 129
bronze bells, 104
brooklime, **167**
brook saxifrage, 162
broomrapes, 165
broomrape family, 98, 165
brooms, 82
brownies, 120
brussel sprouts, 48
buckeye, California, 90, **91**
buckbrush, 150, **151**
buckthorn, 150
buckthorn family, 150–151
buckwheat, 137
   California, 138, **140**
   hot rock, 138
   Lobb's, 138
   naked stem, 138, **140**
   rose, 138
   sulfur, 138, **139**
   wild, 137, 138
buckwheat family, 137–140
budscale, spiny, 61
bulrush, 72
bunchberry, 64
bur-chervil, 25
bur-clover, 82
bursage, 61
   dune, 36
buttercup, California, 147, **149**
   water, 147
buttercup family, 147–149
butterwort, 37
button-bush, western, 156, **157**
button-parsley, 24
button-willow, western, 156, **157**

cabbage, 48
cacao, 108
cacao family, 108
cactus, barrel, 52, **53**
   beavertail, 51
   Christmas, 51
   fishhook, 52
   hedgehog, 52
   mound, 52
   orchid, 51
   velvet, 52
cactus family, 51–53
calabazilla, 68, **69**

calico flower, 135
California holly, 153
California nutmeg, 171
camas, 104
camphor, 101
canaigre, 137, **139**
canary grass tribe, 129
canchalagua, 86
candlenut, 79
candlewood family, 51
candy cane, 77
candy-stripe, 141
caper family, 56
caper plant, 56
caraway, 24
cardinal flower, 54
cardoon, 36
carnation, 59
carrot, 24
carrot family, 24
cascara sagrada, 150
cashew, 22
castor bean, 79
catalina perfume, 89
catclaw, 81, **83**
catchfly, 59
cat's ear, hairy, **38**
cauliflower, 48
cedar, incense, 70, **71**
    Port Orford, 70
    western red-, 70
celery, 24
century plant, 18
cestrum, 168
chaffweed, 143
chamise, 153
chard, 61
chatterbox, 120
checkerbloom, 108, **110**
    mountain, 108
checker-lily, 104
checker-mallow, 108
cheeses, 108
chenille plant, 79
cherry, 152, 153
    holly-leaf, **155**
chestnut, 84
chia, 98
chickweed, 59
    bluff, 59
    meadow, 59
chicory, 34, 35
Chinese houses, 166
chinquapin, coast, 84, **85**
    mountain, 84

chives, 21
cholla, 51
    buckhorn, 52
    coast, 52
    jumping, 52
    teddy bear, 52
chrysanthemum, 34
cinnamon, 101
cinquefoil, shrubby, 152
    sticky, 152, **154**
citrus family, 158–159
clammy-weed, 56
clematis, 147
clerodendron, 156
cliff-rose, 153
clocks, 88
clover, 82
    jackass, 56
cloves, 111
cocklebur, 36
coffee, 156
coffee berry, 150, **151**
coffee family, 156–157
cola, 156
coltsfoot, western, 37
columbine, alpine, 147
    red, 147, **149**
comb-fruit, 46
comfrey, 46
composite family, 34–41
coneflower, 36
copper-leaf, California, 79
coralroot, spotted, 119, **121**
    striped, 119
coriander, 24
corn, 128
corn-lily, 104
cotton, 108
cotton-grass, 72
cottontop, 52
cottonwood, black, 160
    Fremont, 160, **161**
cow-parsnip, 24, **26**
coyote bush, 35
coyote-mint, common, 99, **100**
    mountain, 99
crabapple, native, 153
cranberry, 75
cranesbill, 88
cranesbill family, 88
creambush, 152
cream cups, 123
cream sacs, 166
cress, water, 48
    winter, 48

croton, 79
   California, 79
crowfoot family, 147–149
crown-pink, 49
cucumber, 68
   Indian, 68
cucumber family, 68–69
cudweed, 37
cup-and-saucer vine, 134
currant, chaparral, 89
   golden, 89
   pink-flowering, 89
   red-flowering, 89
   wax, 89
cyclamen, 143
cypress, Lawson, 70
   Monterey, 70
   Sargent, 70, **71**
cypress family, 70–71

dahlia, 34
daisy, 34
   English, 35
   seaside, **38**
daisy family, 34–41
dandelion, common, 35
   native, 35, **38**
date palm, African, 32
   Canary Island, 32
day-lily, 103
death-camas, 104
deerbroom, 82
deerbrush, 150
deer-tongue, 86, **87**
desert-olive, 115
desert-thorn, 168
desert trumpet, 138
dichondra, 63
dill, 24
dock, California, 137
   willow-leaf, 137
dogbane, 28
dogbane family, 28
dogwood, black-fruited, 64
   brown, 64
   mountain, 64
   pacific flowering, 64, **65**
   red-twig, 64, **65**
Douglas-fir, 124, 125, **126**
downingia, 54, **55**
dudleya, 66
   hot rock, 66, **67**

eggplant, 168
elderberry, blue, 57, **58**
   red, 57

elephant snouts, 166
endive, 34
ephedra, green-twig, **74**
escallonia family, 162
eucalypt, 111
everlasting, pearly, 37, **41**
everlasting tribe, 37

fairy bells, 104
fairy slipper, 120
fan palm, California, 32, **33**
   Mexican, 32
farewell-to-spring, 117
fawn-lily, 104
fennel, 24, **26**
fern, bead, 145
   birdsfoot, 146
   California maidenhair, 145
   coffee, 145, **146**
   five-finger, 145, **146**
   goldback, 145
   silverback, 145
   southern maidenhair, 145
fescue, 129
   California, **131**
fetid adder's tongue, 104
fiddleneck, 46
   common, **47**
fiesta flower, 92
fig, Hottentot, 20, **20**
figwort, 166
figwort family, 165–167
filaree, 88, **88**
filbert, 44
fir, grand, 125
   red, 124
   Santa Lucia, 125
   white, 124, **126**
firecracker flower, 173
five-spot, 92
flannel bush, 109
flax family, 88
forget-me-not, 46
flowering-maple, 108
footsteps-to-spring, 24
forsythia, 115
fountaingrass, 130
four o'clock, 113
   desert, **114**
four o'clock family, 113–114
foxglove, 165, 166
foxtail, 129
foxtail tribe, 129
frangipani, 28
fremontia, 109
fringecups, 162

fringepod, 49, **50**
fritillary, 104
frutilla, 168
fumitory, 122
fumitory family, 122

gardenia, 156
garlic, 21
garrya family, 44, 64, 84, 160
gentian, alpine, 86, **87**
    fringed, 86
    green, 86, **87**
    little, 86
    Sierra, 86
    vernal pool, 86
    wanderer's, 86
geranium, 88
    garden, 88
    scented, 88
    wild, 88
ghost pipe, 77
ghost pipe subfamily, 76, 78
giant sequoia, 172
gilia, birdseye, 134, **136**
    globe, 134
ginseng family, 24
glacier-lily, 104
globe flower, 147
globe-tulip, 105
glueseed, 37
goatsbeard, 153
gold dust plant, 64
goldenbushes, 35
golden eardrops, 122
goldenrod, 35
golden stars, 174
golden-yarrow, 37, **40**
goldfields, 36
gooseberry, canyon, 89
    fuchsia-flowered, 89, **89**
    mountain, 89
    Sierra, 89
gooseberry family, 89–90
goosefoot family, 59
gorse, 82
gourd, coyote, 68, **69**
gramma grass tribe, 129
grapefruit, 158
grape-hyacinth, 103
grass, bermuda, 129
    cord, 129
    crab, 130
    dune, 129
    gramma, 129
    hair, 129
    hair, tufted, 129

June, 129
    orchard, 129
    panic, 130
    rabbitsfoot, 128
    rattlesnake, 129
    squirrel-tail, **132**
    vanilla, 129
    velvet, 129
grass family, 128–133
grass-of-Parnassus, 163
green-brier, 104
ground-cherry, 168
ground-ivy, 99
groundsel, 37
    common, **41**
guava, 111
gum, 111
    blue, **112**

hardenbergia, 81
harebell, California, 54
    Scouler's, **55**
hawthorn, 153
hazelnut, 44
    California, 44, **45**
heartsease, western, 177
heather family, 75–78
heather subfamily, 75–77
hebe, 165
hedge-nettle, 98
hedge-parsley, 25
heliotrope, 46
helleborine, 119
hemlock, coast, 125
    mountain, 125
    poison, 24
    water, 24
hemp, Indian, 28, **29**
henbane, 99
hog-fennel, 25, **27**
holly, desert, **62**
hollyhock, 108
    mountain, 108
honesty, 49
honeysuckle, southern, 57
    twinberry, 57, **58**
    vine, 57
honeysuckle family, 57–58
hopbush, 158, **159**
hopsage, 62
horehound, 99
horkelia, 152
hornbeam, 44
horse-chestnut, 90
horse-chestnut family, 90–91
horsemint, 98

COMMON PLANT-NAMES INDEX 197

horse-purslane, 20
hound's tongue, 46
huckleberry, evergreen, 75
   red, 75
hyacinth, 103
hydrangea family, 162

iceplant, 20
iceplant family, 20
Indian paintbrush, Lemmon's, 166
   meadow, 166
   wavy-leaf, 166
   woolly, 166, **167**
Indian warrior, 166
inside-out flower, 42, **43**
iodine bush, 61
iris, coast, 94
   Douglas, 94, **95**
   ground, 94
   mountain, 94
   rainbow, 94
   Sierra, 94
iris family, 94–95
Irish-moss, 59
ironbark, red, 111
ironwood, island, 152
Ithuriel's spear, 173, 174
ivesia, 152
ivy, redwood, 42

Jacob's ladder, 134, **135**
Japanese-cedar, 172
jasmine, 115
jewelflower, 48
   common, **50**
jimson weed, 168
johnny-tuck, 166
joint-fir, 74
joint-fir family, 74
Joshua tree, 18
jujube, Chinese, 150
juniper, California, 70
   mat, 70, **71**
   Sierra, 70
   western, 70

kale, 48
knotweed, Davis's, 137
   dune, **139**
   water, 137
knotweed subfamily, 137
kohlrabi, 48

Labrador tea, 76
lacepod, 49
ladies' tresses, 119
ladyslipper, California, 120, **121**
   mountain, 120
lady's pocket book, 165
lady's thumb, 137
lamb's quarters, 61
lantana, 175
larkspur, 147
   scarlet, 148
   yellow, 148
laurel, bog, 76
laurel family, 101–102
lavender, 98
lemon, 158
lemon balm, 99
lettuce, 34
lewisia, 141
   Siskiyou, 141
licorice, wild, 82
lilac, 115
lilly-pilly tree, 111
lily, 104
   leopard, **106**
lily family, 103–106
lily-of-the-valley family, 16
lily-of-the-valley shrub, 75
lime, 158
linanthus, small-flowered, **136**
live-forever family, 66–67
lizard tail, 37
lobelia, Dunn's, 54
loco-weed, 82
locust, black, 82
loosestrife, 143
lotus, 82
   bull, 82
lousewort, 166
   mountain, 166
lovage, 25
lungwort, 46
lupine, 81, 82
   blue bush, **83**
   false, 82
   silver-leaf, **83**

madder, field, 156
madder family, 156–157
madrone, 75
magic plant, 134
maguey, 16
mallow, alkali, 109
   apricot, 108

bush, 108, 109
globe, 109
Hall's, **110**
marsh, 109
mallow family, 108–110
malva rosa, 108
mango, 22
mangrove, black, 175
manioc, 79
manroot, big-fruited, 68
    common, 68, **69**
    Oregon, 68
manzanita, 75
    common, **77**
maple, bigleaf, 16, **17**
    mountain, 16
    Sierra, 16
    sugar, 16
    vine, 16, **17**
maple family, 16–17
marguerite, 34
marigold, 34
mariposa-tulip, 105
    superb, **106**
marshmallow, 108
marvel of Peru, 113
mayweed, 35
mayweed tribe, 34, 35
meadow beauty, 92
meadowfoam family, 88
meadowrue, 147
melic, 129
melon, 68
    coyote, 68
milkmaids, 48
milk-thistle, 35
milk-vetches, 82
milkweed, California, 28, **30**
    heart-leaf, 28
    showy, 29, **30**
    whorled, 28
millet, 128
miner's lettuce, 141, **142**
mint, 98
    hummingbird, 99
    spear, 99
    vernal pool, 99
mint family, 98–100
mission bells, 104
mist-maidens, 92
mimosa subfamily, 81
mitrewort, 162
mock-orange, Mexican, 158
mock-orange family, 162
monkeyflower, 165
    golden, **167**
    seep, **167**
monkshood, 147
montbretia, 94
montia, little-leaf, 141
Mormon tea, 74
    green-twig, **74**
morning glory, 63
    beach, **63**
    wild, 63
morning glory family, 63
mountain-mahogany, 152, **154**
mousetails, 148, 152
muhly, 128
mulefat, 35, 62
mule's ears, 36
    woolly, **39**
mullein, 165
    turkey, 79, **80**
myrtle, common, 111
myrtle family, 111–112
myrtlewood, 101

needlegrass, 128
    purple, **130**
nightshade, black, 168
nightshade family, 168–169
ninebark, 153
nut-grass, 72, **73**

oak, blue, 84
    California black, 84
    canyon live, 85
    coast live, 85, **85**
    common scrub, 84
    Engelmann, 84
    Garry, 84
    goldcup, 85
    interior live, 84
    scrub, 84
    valley, 84
oat-grass, 129
oats, 128
    wild, 129
ocean spray, 152
okra, 108
oleander, 28
olive, 115
olive family, 115
onion, coast
    crinkled, 21, **21**
    paper, 21
    red-skinned, 21

onion, coast (*continued*)
    swamp, 21
    "wild", 21
onion family, 21
orange, 158
orchid, brook, 120
    ghost, 120
    phantom, 120
    rattlesnake, 119
orchid family, 119–121
oregano, 98
oso-berry, 153
our lord's candle, 18, **19**
owl's clover, 166
oxalis family, 88

palm family, 32–33
panic, hot springs, **133**
panic tribe, 130
pansy, 177
    wild, 177
paperbark, 111
parsley, 24
    water, 25
parsley family, 24–27
parsnip, 24
pasque flower, western, 147
pea, chaparral, 82
pea family, 81–83
pea subfamily, 81–83
peach, 152
pear, 152
pearlwort, 60
pelican beak, 166
pennyroyal, 99
    western, 98
penstemon, 166
    blue foothill, **167**
    shrub, 166
pepper, bell, 168
    chile, 168
pepper grass, 49, **50**
pepper-tree, Chilean, 22
pepperwood, 101, **102**
periwinkle, 29, **31**
petunia, 168
    wild, 168
peyote, 51
phacelia, 92, 93
    rock, **93**
phlox, annual, 135
    mat, 134
    showy, 134
pickleweed, 51

piggyback plant, 162
pincushion flower, 37
    yellow, **40**
pine, bishop, 124
    bristlecone, 124
    foothill, 124
    gray, 124
    Jeffrey, 124
    ponderosa, 124, **126**
    white, 124
pinedrops, 76
pine family, 124–126
pink, Hooker's, 59
    Indian, 59, **60**
    native, 59
    windmill, 59
pipsissewa, 76, **78**
pistache, Chinese, 22
pistachio, 22
pitcher-sage, 98
plane tree, London, 127
plane tree family, 127
plum, 152
    wild, 153
podocarp family, 170
poinsettia, 79
poison-ivy, 22
poison-oak, 22, **23**
polypody family, 145
poodle dog plant, 92
popcorn flowers, 46
poppy, blue, 122
    bush, 122
    California, 122, **123**
    fire, 123
    flame, 123
    Matilija, 122, **123**
    opium, 122
    prickly, 123
    wind, 123
poppy family, 122–123
portulaca, 141
portulaca family, 141–142
potato, 168
potato family, 168–169
poverty weed, 36
prickly ox-tongue, 35
prickly-pear, coast, 52
    Mojave, 51, **53**
    old man, 51, **53**
prickly-phlox, 134
Pride-of-Madeira, 46
primrose, 143
    Sierra, 143, **144**
primrose family, 143–144

prince's pine, 76, **78**
prince's plume, 56
puccoon, 46
pumpkin, 68
purslane, 141
purslane family, 141–142
pussy paws, 141
pussy toes, 37
pygmy weed, 66
pyrola, rose, 76

quaking aspen, 160
Queen Ann's lace, 25
quinoa, 61

radicchio, 34
radish, wild, 48
ragweed, 36
ragweed subtribe, 36
ranger's buttons, 24
ranunculus, 147
raspberry, 152, 153
rattlepod, 82
    speckled, **83**
redberry, holly-leaf, 150, **151**
redbud, western, 81, **83**
red-heather, 76
red maids, 141
red-shanks, 153
redwood, coast, 172, **172**
    dawn, 172
    Sierra, 172
redwood family, 172
reed, giant, 129
    native, 129
rein orchid, green, 119
    snowy, 119
rhododendron, 75
rhubarb, 137
    Indian, 137, **139**, 162
rice, 128
rockcress, 48
    coast, **50**
rock-jasmine, 143
rock-nettle, 107
rock-spiraea, 153
rose, 152
    California, **155**
    wild, 153
rose family, 152–155
rosebay, 75
    California, 76
rosecrown sedum, 66, **67**

rosilla, **40**
rosinweed, 36
rue-anemone, 147
rue family, 158–159
rush, 94, 96, **97**
    toad, **97**
rush family, 96–97
Russian-thistle, 61
rutabaga, 48
rye-grasses, European, 130
    native, 130

sage, 98
    black, 98, **100**
    hummingbird, 98
    purple, 98
    thistle, 98
    white, 98
sagebrush, 35
saguaro, giant, 52, **53**
salal, 75
salmonberry, 153
saltbush, 61
sand paper plant, 107
sand-spurrey, 59
sand-verbena, desert, 113
    pink, 113, **114**
    red, 113
    yellow, 113
sandwort, 59
    King's, **60**
sanicle, poison, 24
    purple, 24
    woodland, 24, **26**
sassafras, 101
savory, native, 98
saxifrage, 162
    golden, 163
saxifrage family, 162–164
scallions, 21
scarlet-gilia, 134
scarlet pimpernel, 143
sea-blite, 61
sea-grape, 137
sea-milkwort, 143
sea-purslane, western, 20
sedge, 72
    southern, **73**
    umbrella, 72
sedge family, 72–73
self-heal, 99
senecio, bush, **41**
senecio tribe, 37
senna subfamily, 81

service berry, 153
shallots, 21
shepherd's needles, 25
shepherd's purse, 49
shield-leaf, **50**
shinleaf, white-veined, 76
shooting stars, Henderson's, 143, **144**
    padre's, 143
sisal, 18
silver puffs, 35
skullcap, 99
slink pods, 103
smoketree, 22
snapdragon, 165
    island, 166
snapdragon family, 165–167
sneezeweed, 37
sneezeweed subtribe, 36, 37
snowberry, 57
    shrub, **58**
snow-in-summer, 59
snowplant, 77, **78**
soapberry family, 16, 90
soap plant, 104
soaproot, 61
soapwort, 60
solomon's seal, false, 104
    fat, **105**
sorghum, 128
sorrel, French, 137
    mountain, 137
sourberry, 22
sow thistle, 35
Spanish dagger, 18
spearpod, 49
speedwell, **167**
spiderflower, 56
    little, 56
spiderling, 113
spike-rush, 72
spineflowers, 138
spiraea, 152
spleenwort family, 145
spring beauty, 141
spruce, Sitka, 125
spurge, 79
    lesser, **80**
    prostrate, 79
spurge family, 79–80
spurrey, 60
squash, 68
squawbush, 22
stapeliad, 28
star-jasmine, 28
star-lily, 104

star-thistle, 36
star-tulip, 105
steeple bush, 152
stinking-yew, 170
stonecrop, 66
    annual, 66
    common, 66, **67**
stonecrop family, 66–67
strawberry, 152
    mock, 152
strawberry tree, 75
stringy bark, 111
sugar bush, 22, **23**
sugar cane, 128
sugar-scoops, 162
sugarsticks, 77
sumac family, 22–23
summer-holly, 75
sunbonnets, 135
sunflower, 34, 36
    little, 36
sunflower family, 34–41
sunflower tribe, 36
sweet-after-death, 42
sweet cicely, 25
sweet-clover, 81, 82
sweet pea, 81
    wild, 82
sweet potato, 63
sycamore, western, 127, **127**

tanbark-oak, 84
tangerine, 158
tansy, 34
tarplants subtribe, 36
tarragon, 34
tarweed, elegant, **39**
tauschia, 25
teak, 175
tear-thumb, 137
tea tree, Australian, 111
thimbleberry, 153
thistle, bull, 36, **39**
    cobweb, 36, **39**
    plumeless, 36
thistle tribe, 35–36
thorn-apple, 168, **169**
thyme, 98
tidy tips, 36
timothy, 128
tincture plant, 166
toadflax, 165
    native, 166
tobacco, tree, 168
    wild, 168

tomatillo, 168
tomato, 168
toyon, 153
trail plant, 37
trillium, 104
trumpet-vine family, 98
tule, 72
tulip, 103
tumble weed, 61
tunic flower, 60
turnip, 48
turpentine broom, 158
twinflower, 57
twisted stalk, 104

umbrella-pine, 172
umbrella plant, 162

vanilla, 119
vanilla leaf, 42
venus' looking glass, 54
verbena, delta, 175, **176**
    desert, 175
    lemon, 175
veronica, 165
vervain, common, 175
vetch, common, 82
    giant, 82
viburnum, native, 57
vine family, 68
violet, dog, 177
    Douglas's, 177
    English, 177
    lobed, 177
    pine, 177
    redwood, 177, **178**
    Shelton's, 177
    smooth yellow, 177
    white meadow, 177
    wild, 177
virgin's bower, 147

wallflower, 48
    western, 48, **50**

wand flower, 35
waterleaf, 92
    western, 92
waterleaf family, 92–93
watermelon, 68
water-primrose, 116
wax plant, 28
wheat, 128
whispering bells, 93
white-heather, 76
whitlow grass, 49
wild lilacs, 150
willow, alpine, 160
    arroyo, 160, **161**
    snow, 160
willow family, 160–161
wind flower, 147
wintergreen, one-sided, 76
    wild, 76
wisteria, 81
witch-hazel family, 127
wood fern family, 145
woodland nymph, 76
woodland star, 162, **164**
woodmint, 98
    common, **100**
woodrush, 96
woolly blue-curls, 98
woolly heads, 24, 134
woolly star, 135
wormwood, 34, 61

yampah, 25
yarrow, 34
yellow-eyed grass, 94
yerba buena, 98
yerba mansa, 34
yerba santa, 92, **93**
yesterday-today-and-tomorrow, 168
yew, pacific, 170
    western, 170
yew family, 170–171
yuca, 79
yucca, chaparral, 18, **19**

zinnia, 34

# SCIENTIFIC PLANT-NAMES INDEX

Bolded numbers indicate pages with figures.

*Abelia*, 57
*Abies bracteata*, 124
    *concolor*, 124, **126**
    *grandis*, 125
    *magnifica*, 124
*Abronia latifolia*, 113
    *maritima*, 113
    *umbellata*, 113, **114**
    *villosa*, 113
*Abutilon*, 108
*Acacia greggii*, 81, **83**
*Acalypha californica*, 79
*Acanthomintha*, 99
Acanthaceae, 98, 165
*Acer circinatum*, 16, **17**
    *glabrum*, 16
    *macrophyllum*, 16, **17**
    *negundo californicum*, 16, **17**
    *saccharum*, 16
*Achillea millefolium*, 34
*Achlys*, 42
*Achnatherum*, 128
*Achyrachaena mollis*, 36
*Acmena smithii*, 111
*Aconitum columbianum*, 148
*Actaea arguta*, 148
*Adenocaulon bicolor*, 37
*Adenostoma fasiculatum*, 153
*Adiantum aleuticum*, 145, **146**
    *capillus-veneris*, 145
    *jordanii*, 145
Adoxaceae, 57
*Aeonium*, 66
*Aesculus californica*, 90, **91**

    *parryi*, 90
Agavaceae, 18
*Agave deserti*, 18
    *schidigera*, 18
    *sisalana*, 18
    *utahensis*, 18
*Agoseris*, 35, **38**
Agrostideae, 128
*Aira caryophylla*, 129
Aizoaceae, 20
*Alcea rosea*, 108
*Aleurites moluccana*, 79
*Allenrolfea occidentalis*, 61
Alliaceae, 21
*Allium amplectens*, 21
    *crispum*, 21, **21**
    *dichlamydeum*, 21
    *haematochiton*, 21
    *triquetrum*, 21
    *validum*, 21
*Allophyllum*, 134
*Alnus incana tenuifolia*, 44, **45**
    *rhombifolia*, 44
    *rubra*, 44
*Aloe vera*, 103
*Aloysia citriodora*, 175
Amaranthaceae, 61, 137
Amaryllidaceae, 21, 173
*Ambrosia chamissonis*, 36
Ambrosiinae, 36
*Amelanchier*, 153
*Ammophila arenaria*, 128
*Amsinckia menziesii*, 46, **47**
Anacardiaceae, 22–23

*Anacardium excelsum*, 22
*Anagallis arvensis*, 143
*Anaphalis margaritacea*, 37, **41**
*Androsace*, 143
*Androstephium breviflorum*, 173
*Anemone occidentalis*, 147
*Anemopsis californica*, 34
*Angelica*, 24
*Antennaria*, 37
*Anthemideae*, 34
*Anthemis cotula*, 35
*Anthriscus caucalis*, 25
*Antirrhinum*, 166
Apiaceae, 24–27
Apocynaceae, 28
*Apocynum cannabinum*, 28, **29**
*Aquilegia formosa*, 147, **149**
　*pubescens*, 147
*Arabis blepharophylla*, 48, **50**
Araliaceae, 24
*Arbutus menziesii*, 75
　*unedo*, 75
*Arctostaphylos*, 75
　*manzanita*, **77**
Arecaceae, 32–33
*Arenaria kingii*, 59, **60**
*Argemone*, 123
Aristolochiaceae, 42, 101
*Arnica*, 37
*Artemisia spinescens*, 35, 61
*Aruncus vulgaris*, 153
*Arundo donax*, 129
Asclepiadaceae, 28
*Asclepias californica*, 28, **30**
　*cordifolia*, 28
　*fascicularis*, 28
　*speciosa*, 29, **30**
*Asparagus officinalis*, 103
Asphodelaceae, 103
Aspleniaceae, 145
*Aster*, 35
Asteraceae, 34–41
Astereae, 35, **38**
*Astragalus*, 82
　*whitneyi*, **83**
*Atriplex hymeneletra*, 61, **62**
*Aucuba japonica*, 64
*Aurinia saxatilis*, 48
*Avena*, 129
Aveneae, 129
*Avicennia marina resinifera*, 175

*Babiana*, 94
*Baccharis*, 35

*Balsamorhiza*, 36
*Barbarea orthoceras*, 48
*Bellardia trixago*, 165
*Bellis perennis*, 35
Berberidaceae, 42–43
*Berberis nervosa*, 42, **43**
*Bergerocactus emoryi*, 52
*Bernardia myriophylla*, 79
*Beta vulgaris*, 61
*Betula glandulosa*, 44
　*occidentalis*, 44
Betulaceae, 44–45
Bignoniaceae, 98, 165
Blechnaceae, 145
*Blennosperma nanum*, 37
*Bloomeria crocea*, 174
*Boerhavia*, 113
*Bolandra californica*, 162
Boraginaceae, 46–47
*Borago officinalis*, 46
*Bouteloua*, 129
*Boykinia*, 162
*Brassica*, 48
　*oleracea*, 48
　*rapa*, **50**
Brassicaceae, 48–50
*Brodiaea californica*, 173
　*elegans*, 173
　*terrestris*, 173
*Bromus*, 129
*Brugmansia*, 168
*Brunfelsia*, 168


Cactaceae, 51–53
*Calamagrostis*, 128
*Calandrinia ciliata*, 141
*Calceolaria*, 165
*Callistemon*, 111
*Callitropsis*, 70
*Calocedrus decurrens*, 70, **71**
*Calochortus albus*, 105
　*amabilis*, 105
　*amoenus*, 105
　*coeruleus*, 105
　*monophyllus*, 105
　*superbus*, **106**
　*tolmiei*, 105
　*uniflorus*, 105
*Calycadenia*, 36
*Calypso bulbosa*, 120
*Calyptridium umbellatum*, 141
*Calystegia soldanella*, **63**
*Camassia quamash*, 104
*Camissonia boothii*, 116

*cheiranthifolia*, 116
  *ovata*, 116
*Campanula prenanthoides*, 54
  *rotundifolia*, 54
  *scouleri*, **55**
*Cantua*, 134
*Capparaceae*, 56
*Capparis spinosa*, 56
*Caprifoliaceae*, 57–58
*Capsella bursa-pastoris*, 49
*Cardamine californica*, 48
*Carduus*, 36
*Carex spissa*, 72, **73**
*Carnegia gigantea*, 52, **53**
*Carpinus*, 44
*Carpobrotus chilensis*, 20
  *edulis*, **20**
*Caryophyllaceae*, 59–60
*Cassiope mertensiana*, 76
*Castanea*, 84
*Castilleja*, 165
  *applegatei*, 166
  *densiflora*, 166
  *exserta*, 166
  *foliolosa*, 166, **167**
  *lemmonii*, 166
  *miniata*, 166
  *rubicundula*, 166
*Catharanthus*, 28
*Cattleya*, 119
*Caulanthus*, 48
*Ceanothus*, 150
  *cuneatus*, **151**
  *integerrimus*, **151**
*Centaurea*, 36
*Centaurium*, 86
*Centunculus minimus*, 143
*Cephalanthera austinae*, 120
*Cephalanthus occidentalis*, 156, **157**
*Cerastes*, 150
*Cerastium arvense*, 59
  *tomentosum*, 59
*Cercis occidentalis*, 81, **83**
*Cercocarpus betuloides*, 153, **154**
*Ceropegia*, 28
*Chaenactis glabriuscula*, 37, **40**
*Chamaecyparis lawsoniana*, 70
*Chamaesaracha nana*, 168
*Chamaesyce*, 79
*Chamomilla suaveolens*, 35
*Cheilanthes*, 145
*Cheiranthus*, 48
*Chenopodiaceae*, 59, 61–62
*Chenopodium californicum*, 61
  *quinoa*, 61

*Chimaphila*, 76, **78**
*Chlorideae*, 129
*Choisya ternata*, 158
*Chorizanthe*, 138
*Chrysolepis chrysophylla*, 84, **85**
  *sempervirens*, 84
*Chrysosplenium glechomifolia*, 163
*Cicendia quadrangularis*, 86
*Cichorieae*, 35
*Cichorium intybus*, 35
*Cicuta douglasii*, 24
*Cinnamomum camphoratum*, 101
  *zeylandica*, 101
*Circaea alpina*, 116
*Cirsium occidentale*, 36, **39**
  *vulgare*, 36, **39**
*Citrus*, 158
*Clarkia amoena*, 117
  *biloba*, 117
  *breweri*, 117
  *concinna*, 117
  *gracilis*, 117
  *purpurea*, 117
  *rhomboidea*, 117
  *unguiculata*, 117, **118**
*Claytonia gypsophiloides*, 141
  *perfoliata*, 141, **142**
  *sibirica*, 141
*Cleomaceae*, 56
*Cleome*, 56
*Cleomella*, 56
*Clintonia*, 104
  *andrewsiana*, 104
  *uniflora*, 104
*Cneoridium dumosum*, 158
*Cobaea scandens*, 134
*Coccoloba prolifera*, 137
*Codiaeum*, 79
*Coffea arabica*, 156
*Cola nitida*, 156
*Coleonema*, 158
*Collinsia heterophylla*, 166
  *tinctoria*, 166
  *torreyi*, 166
*Collomia grandiflora*, 134
*Comarostaphylis diversifolia*, 75
*Conium maculatum*, 25
*Conocosia pugioniformis*, 20
*Convallariaceae*, 18
*Convolvulaceae*, 63, 168
*Convolvulus*, 63
*Corallorhiza maculata*, 119, **121**
  *striata*, 119
*Cordia*, 46
*Cordylanthus*, 166

*Cordyline*, 32
*Coreopsis*, 36
Cornaceae, 64–65
*Cornus canadensis*, 64
    *glabrata*, 64
    *nuttallii*, 64, **65**
    *sericea*, 64, **65**
    *sessilis*, 64
*Corokia cotoneaster*, 64
*Cortaderia*, 129
*Corydalis caseana*, 122
*Corylus cornuta californica*, 44, **45**
*Cotinus coggygria*, 22
*Crassula*, 66
Crassulaceae, 66–67
*Crataegus*, 153
*Crocosmia*, 94
*Croton californica*, 79
*Cryptantha*, 46
*Cucurbita foetidissima*, 68, **69**
Cucurbitaceae, 68–69
Cupressaceae, 70–71
*Cupressus macrocarpa*, 70
    *sargentii*, 70, **71**
*Cycladenia humilis*, 29
*Cylindropuntia acanthocarpa*, 52
    *bigelovii*, 52
*Cymbidium*, 119
*Cynara cardunculus*, 36
*Cynareae*, 35–36
*Cynodon dactylon*, 129
*Cynoglossum*, 46
Cyperaceae, 72–73
*Cyperus*, 72
    *esculentus*, **73**
*Cypripedium californicum*, 120, **121**
    *fasciculatum*, 120
    *montanum*, 120
*Cytisus*, 82

*Dactylis glomeratus*, 129
*Danthonia*, 129
*Darmera peltata*, 162
*Datura wrightii*, 168, **169**
*Daucus carota*, 25
*Delphinium cardinale*, 148
    *gypsophilum pallescens*, 148
    *hesperium*, 148
    *luteum*, 148
    *nudicaule*, 148
    *purpusii*, 148
*Dendromecon*, 122
*Dendrobium*, 119

Dennstaediaceae, 145
*Deschampsia*, 129
*Dianthus caryophylla*, 59
*Dicentra formosa*, 122
    *chrysantha*, 122
*Dichelostemma capitatum*, 173, **174**
    *ida-maia*, 173
    *volubile*, 173
*Dichondra*, 63
*Dicoria*, 36
*Digitalis purpurea*, 166
*Digitaria*, 130
*Disporum*, 104
*Distichlis spicata*, 129, **132**
*Ditaxis*, 80
*Dodecatheon clevelandii*, 143
    *hendersonii*, 143, **144**
*Downingia concolor*, 54, **55**
*Draba*, 49
*Dracaena*, 32
*Draperia systyla*, 92
Dryopteridaceae, 145
*Duchesnea indica*, 152
*Dudleya cymosa*, 66, **67**
    *farinosa*, 66

*Echeveria*, 66
*Echinocactus polycephalus*, 52
*Echinocereus engelmannii*, 52
    *triglochidiatus*, 52
*Echium*, 46
*Eleocharis*, 72
*Elymus elymoides*, 130, **132**
*Emmenanthe penduliflora*, 93
*Ephedra viridis*, **74**
Ephedraceae, 74
*Epidendrum*, 119
*Epilobium angustifolium*, 116
    *canum*, 116, **118**
    *densifolia*, 116
    *obcordatum*, 116
*Epimedium*, 42
*Epipactis gigantea*, 120
    *helleborine*, 120
*Epiphyllum*, 51
*Eremocarpus setigerus*, 79, **80**
*Eriastrum*, 135
*Erica*, 75
Ericaceae, Ericoideae, 75–77
    Monotropoideae, 76, 78
*Ericameria*, 35
*Erigeron*, 35
    *glaucus*, **38**

*Eriodictyon californicum*, 92, **93**
*Eriogonoideae*, 137, 138
*Eriogonum compositum*, 138
    *fasciculatum*, 138, **140**
    *inflatum*, 138
    *lobbii*, 138
    *luteolum*, 138
    *nudum*, 138, **140**
    *spergulinum*, 138
    *umbellatum*, 138, **139**
*Eriophorum gracile*, 72
*Eriophyllum*, 37
    *confertiflorum*, **40**
*Erodium brachycarpum*, 88
    *cicutarium*, **88**
*Eryngium*, 24
*Erysimum capitatum*, 48, **50**
*Erythronium*, 104
*Escalloniaceae*, 162
*Eschscholzia californica*, 123, **123**
*Escobaria*, 52
*Eucalyptus globulus*, 111, **112**
    *sideroxylon*, 111
*Eucnide rupestris*, 107
    *urens*, 107
*Eucrypta*, 92
*Euphorbia peplus*, **80**
    *pulcherrima*, 79
Euphorbiaceae, 79–80

Fabaceae, Caesalpinoideae, 81
    Mimosoideae, 81
    Papillioinoideae, 81–83
Fagaceae, 84–85
*Fagopyrum esculentum*, 137
*Fagus*, 84
*Ferocactus cylindraceus*, 52, **53**
    *viridescens*, 52
*Festuca californica*, 129, **131**
*Festuceae*, 129, **131**
*Foeniculum vulgare*, 25, **27**
*Forestiera pubescens*, 115
*Forsythia*, 115
*Fouquieria splendens*, 51
Fouquieriaceae, 51
*Fragaria*, 152
*Fraxinus dipetala*, 115
    *latifolia*, **115**
    *velutina*, 115
*Freesia retrofracta*, 94, 95
*Fremontodendron*, 109
*Fritillaria*, 104
*Fuchsia*, 116

*Fumaria*, 122
Fumariaceae, 122

*Galium*, 156, **157**
*Galvezia speciosa*, 166
Garryaceae, 44, 64, 84
*Gaultheria shallon*, 75
*Gayophytum*, 116
*Genista*, 82
*Gentiana calycosa*, 86
    *newberryi*, 86, **87**
Gentianaceae, 86–87
*Gentianella amarella*, 86
*Gentianopsis holopetala*, 86
Geraniaceae, 88
*Geranium carolinianum*, 88
    *dissectum*, 88
    *molle*, 88
    *richardsonii*, 88
*Geum*, 152
*Gilia capitata*, 134
    *leptalea*, 134
    *tricolor*, 134, **136**
*Githopsis pulchella*, 54, **55**
*Gladiolus*, 94
*Glaux maritima*, 143
*Glecoma hederacea*, 99
*Glycyrrhiza lepidota*, 82
*Gnaphalium*, 37
*Goodyera oblongifolia*, 119
*Gossypium*, 108
*Grayia spinosa*, 62
*Grindelia*, 35
Grossulariaceae, 89–90

Haemadoraceae, 94
Hammamelidaceae, 127
*Hastingsia alba*, 104
*Hazardia*, 35
*Hebe*, 165
*Helenieae*, 36
*Helenium puberulum*, 37, **40**
*Heliantheae*, 36
*Helianthella*, 36
*Helianthus*, 36
*Heliotropium curassavicum*, 46
*Hemerocallis*, 103
*Heracleum lanatum*, 24, **26**
*Hesperochiron*, 92
*Hesperostipa*, 128
*Hesperoyucca whipplei*, 18, **19**
*Heterocodon rariflorum*, 54
*Heteromeles arbutifolia*, 153

*Heuchera micrantha*, 162, **164**
*Hibiscus esculentus*, 108
   *lasiocarpus*, 109
   *rosa-sinensis*, 108
*Hieracium*, 35
*Hierchloe occidentalis*, 129, **131**
Hippocastanaceae, 90–91
*Holcus lanatus*, 129
*Holodiscus*, 153
Hordeae, 129–130
*Hordeum*, 129
*Horkelia*, 152
*Hoya*, 28
*Hyacinthus*, 103
Hydrangeaceae, 162
Hydrophyllaceae, 92–93
*Hydrophyllum occidentale*, 92
   *tenuipes*, 92
*Hypochaeris*, 35
   *radicata*, **38**

*Iliamna bakeri*, 109
   *latibracteata*, 109
Inuleae, 37
*Ipheion uniflorum*, 21
*Ipomoea batatas*, 63
*Ipomopsis*, 134
Iridaceae, 94–95
*Iris douglasiana*, 94, **95**
   *hartwegii*, 94
   *longipetala*, 94
   *macrosiphon*, 94
   *missouriensis*, 94
*Isocoma*, 35
*Isomeris arborea*, 56, **56**
*Isopyrum*, 147
*Ivesia*, 152

*Jasminum*, 115
*Jepsonia*, 162
Juncaceae, 96–97
*Juncus bufonius*, 96, **97**
*Juniperus californica*, 70
   *communis saxatilis*, 70, **71**
   *occidentalis*, 70

*Kalanchoe*, 66
*Kalmia polifolia microphylla*, 76
*Keckiella*, 166
*Kelloggia galioides*, 156
*Koeleria macrantha*, 129
*Krascheninnikovia lanata*, 62

Lamiaceae, 98–100
*Lamium amplexicaule*, 99
*Langloisia*, 135
*Lantana camara*, 175
*Lasthenia*, 36
*Lathyrus*, 82
Lauraceae, 101–102
*Laurus nobilis*, 101
*Lavatera assurgentiflora*, 108
*Layia*, 36
*Ledum glandulosum*, 76
*Lepechinia*, 98
*Lepidium*, 49, **50**
*Leptodactylon californicum*, 135
   *pungens*, 135
*Leptospermum laevigatum*, 111
*Lesquerella*, 49
*Lessingia*, 35
*Lewisia cotyledon*, 141
   *rediviva*, 141, **142**
*Leymus*, 130
*Ligusticum apiifolium*, 25
   *grayi*, 25
Liliaceae, 103–106
*Lilium pardalinum*, 104, **106**
Limnanthaceae, 88
Linaceae, 88
*Linanthus*, 135
   *parviflorus*, **136**
*Linnaea borealis*, 57
*Lithophragma*, 162, **164**
*Lithospermum*, 46
Loasaceae, 107
*Lobelia cardinalis*, 54
   *dunnii*, 54
Lobeliaceae, 54
*Lobularia maritima*, 49
*Loeseliastrum*, 135
*Lolium*, 130
*Lomatium dasycarpum*, 25, **27**
*Lonicera ciliosa*, 57
   *hispidula vacillans*, 57
   *interrupta*, 57
   *involucrata*, 57, **58**
   *subspicata*, 57
*Lotus crassifolius*, 82
   *scoparius*, 82
*Ludwigia*, 116
*Luma apiculata*, 111
*Lunaria annua*, 49
*Lupinus*, 82
   *albifrons*, **83**
*Luzula*, 96
*Lychnis coronaria*, 59
*Lycianthes*, 168

*Lycium*, 168
*Lyonothamnus floribundus*, 152
*Lysimachia nummularia*, 143
   *thyrsiflora*, 143

*Madia elegans*, **39**
*Madiinae*, 36, **39**
*Mahonia*, 42
*Malacothamnus fasciculatus*, 109
   *hallii*, 109, **110**
*Malosma laurina*, 22
*Malus fusca*, 153
*Malva*, 108
Malvaceae, 108–110
*Malvella leprosa*, 109
*Mamillaria*, 52
*Mangifera indica*, 22
*Manihot esculenta*, 79
*Marah fabaceus*, 68, **69**
   *macrocarpus*, 68
   *oreganus*, 68
*Marrubium vulgare*, 99
*Matthiola*, 48
*Meconella*, 123
*Meconopsis*, 122
*Medicago*, 82
*Melaleuca*, 111
*Melica*, 129
*Melilotus*, 82
*Melissa officinalis*, 99
*Menodora*, 115
*Mentha arvensis*, 99
   *pulegium*, 99
   *spicata*, 99
*Mentzelia involucrata*, 107
   *laevicaulis*, 107, **107**
   *lindleyi*, 107
*Mertensia ciliata*, 46, **47**
*Mesembryanthemum crystallinum*, 20
*Microseris*, 35
*Mimulus*, 166
   *guttatus*, **167**
*Minuartia*, 59
*Mirabilis multiflora*, 113, **114**
*Mitella*, 162
*Monardella macrantha*, 99
   *odoratissima*, 99
   *villosa*, 99, **100**
*Moneses uniflora*, 76
*Monolopia*, 37
*Monotropa*, 77
*Montia parvifolia*, 141
*Muhlenbergia*, 128
*Muilla maritima*, 174

*Muscari*, 103
*Myosotis*, 46
*Myosurus*, 148
Myrtaceae, 111
*Myrtus communis*, 111

*Nandina domestica*, 42
Nartheciaceae, 104
*Narthecium californicum*, 104
*Nassella*, 128
   *pulchra*, **130**
*Navarretia*, 134
*Nemacladus*, 54
*Nemophila maculata*, 92
   *menziesii*, 92
*Nerium oleander*, 28
*Nicotiana glauca*, 168
*Nolina*, 18
*Nothofagus*, 84
Nyctaginaceae, 113–114

*Oemleria cerasiformis*, 153
*Oenanthe sarmentosa*, 25
*Oenothera*, 116
   *deltoides*, 117
   *elata hookeri*, 117, **118**
*Olea europea*, 115
Oleaceae, 115
*Omphalodes*, 46
Onagraceae, 116–118
*Oncidium*, 119
*Opuntia chlorotica*, 51, **53**
   *erinacea*, 51, **53**
   *ficus-indica*, 51
   *littoralis*, 52
Orchidaceae, 119–121
Orobanchaceae, 98, 165
*Orthilia secunda*, 76
*Orthocarpus*, 166
Oxalidaceae, 88
*Oxyria digyna*, 137

Paniceae, 130
*Panicum thermale*, 130, 133
*Papaver californicum*, 122–123
Papaveraceae, 122–123
*Paphiopedilum*, 119
*Parentucelia viscosa*, 165
*Parnassia*, 163
*Parvisedum*, 66
*Pedicularis densiflora*, 166
   *groenlandica*, 166
   *semibarbata*, 166

*Pediomelum californicum*, 82
*Pelargonium*, 88
*Pellaea andromedifolia*, 145, **146**
    *mucronata*, 146
*Pennisetum*, 130
*Penstemon heterophyllus*, 166, **167**
*Pentagramma triangularis*, 145
*Perideridia*, 25
*Persea americana*, 101
*Petalonyx thurberi*, 107
*Petasites frigidus palmatus*, 37
*Petrophyton caespitosum*, 153
*Petroraghia prolifera*, 60
*Petunia parviflora*, 168
*Phacelia imbricata*, **93**
*Phalaenopsis*, 119
*Phalarideae*, 129
Philadelphaceae, 162
*Phleum*, 128
*Phlox diffusa*, 134
    *gracilis*, 135
    *speciosa*, 134
*Phoenicaulis cheiranthoides*, 49
*Phoenix canariensis*, 32
    *dactylifera*, 32
*Pholistoma auritum*, 92
*Phragmites australis*, 129
Phrymaceae, 165
*Phyla nodiflora*, 175
*Phyllodoce*, 76
*Physalis*, 168
*Physocarpus capitatus*, 153
*Picea sitchensis*, 125
*Pickeringia montana*, 82
*Picris echioides*, 35
*Pieris japonica*, 75
*Pimenta dioica*, 111
Pinaceae, 124–126
*Pinus contorta murrayana*, 124
    *jeffreyi*, 124
    *lambertiana*, 124
    *longaeva*, 124
    *muricata*, 124
    *ponderosa*, 124, **126**
    *radiata*, 124
    *sabiniana*, 124
*Piperia*, 119
*Pistacia chinensis*, 22
    *vera*, 22
Plantaginaceae, 165
Platanaceae, 127
*Platanus racemosa*, **127**
    x *acerifolia*, 127
*Platanthera leucostachys*, 119
    *sparsiflora*, 119

*Platystemon californicus*, 123
*Plumeria rubra*, 28
*Poa*, 129
Poaceae, 128–133
Podocarpaceae, 170
*Pogogyne*, 99
*Polanisia dodecandra*, 56
Polemoniaceae, 134–136
*Polemonium occidentale*, 134, **135**
Polygonaceae, 137–140
*Polygonum amphibium*, 137
    *bistortoides*, 137
    *davsiae*, 137
    *paronychia*, 137, **139**
    *phytyloccaefolium*, 137
Polypodiaceae, 145
*Polypogon monspeliensis*, 128
*Populus balsamifera trichocarpa*, 160
    *fremontii*, 160, **161**
    *tremuloides*, 160
*Portulaca oleracea*, 141
Portulacaceae, 141–142
*Potentilla fruticosa*, 152
    *glandulosa*, 152, **154**
*Primula suffrutescens*, 143, **144**
Primulaceae, 143–144
*Prunella vulgaris*, 99
    var. *lanceolata*, 99
*Prunus ilicifolia*, 153, **155**
*Pseudotsuga menziesii*, 125, **126**
*Ptelea crenulata*, 158, **159**
Pteridaceae, 145–146
*Pterospora andromeda*, 76
*Pterostegia drymarioides*, 138
*Pulmonaria*, 46
*Purshia*, 153
*Pyrola asarifolia*, 76
    *picta*, 76
Pyrolaceae, 75


*Quercus agrifolia*, **85**
    *berberidifolia*, 84
    *chrysolepis*, 85
    *douglasii*, 84
    *engelmannii*, 84
    *garryana*, 84
    *kelloggii*, 84
    *lobata*, 84
    *vaccinifolia*, 85
    *wislizenii*, 85


*Raillardella*, 37
Ranunculaceae, 147–149

*Ranunculus aquatilis*, 147
    *californicus*, 147, **149**
*Raphanus*, 48
Rhamnaceae, 150–151
*Rhamnus californica*, 150, **151**
    *ilicifolia*, 150, **151**
    *purshianus*, 150
*Rhododendron*, 75
    *macrophyllum*, 76
    *occidentale*, 76, **77**
*Rhus integrifolia*, 22
    *ovata*, 22, **23**
    *trilobata*, 22
*Ribes aureum*, 89
    *cereum*, 89
    *malvaceum*, 89
    *menziesii*, 89
    *montigenum*, 89
    *roezlii*, 89
    *sanguineum*, 89
    *speciosum*, 89, **89**
    *viburnifolium*, 89
*Ricinus communis*, 79
*Robinia pseudo-acacia*, 82
*Romanzoffia*, 93
*Romneya*, 122, **123**
*Rorippa nasturtium-aquaticum*, 48
*Rosa californica*, 153, **155**
Rosaceae, 152–156
Rubiaceae, 156–157
*Rubus*, 153
*Rudbeckia californica*, 36
*Rumex acetosa*, 137
    *californicus*, 137
    *hymenosepalus*, 137, **139**
    *salicifolius*, 137
Rutaceae, 158–159

*Sagina*, 59
Salicaceae, 160–161
*Salix arctica*, 160
    *lasiolepis*, 160, **161**
    *nivea*, 160
*Salsola tragus*, 61
*Salvia apiana*, 98
    *carduacea*, 98
    *columbariae*, 98
    *leucophylla*, 98
    *mellifera*, 98, **100**
    *spathacea*, 98
*Sambucus mexicana*, 57, **58**
    *racemosa*, 57
*Sanicula arctopoides*, 24
    *bipinnata*, 24
    *bipinnatifida*, 24
    *crassicaulis*, 24, **26**
Sapindaceae, 16, 89
*Saponaria officinalis*, 60
*Sarcobatus vermiculatus*, 61
*Sarcodes sanguinea*, 77, **78**
*Sassafras albida*, 101
*Satureja mimuloides*, 98
*Saxifraga*, 163
Saxifragaceae, 162–164
*Scandix pecten-aboriginum*, 25
*Schinus molle*, 22
*Scirpus acutus*, 72
    *californicus*, 72
*Scoliopus bigelovii*, 104
*Scrophularia*, 166
Scrophulariaceae, 165–167
*Scutellaria*, 99
*Sedum roseum integrifolium*, 66, **67**
    *spathulifolium*, 66, **67**
*Sempervivum*, 66
*Senecio flaccidus douglasii*, **41**
Senecioneae, 37, **41**
*Sequoia sempervirens*, **172**
*Sequoiadendron giganteum*, 172
*Sesuvium verrucosum*, 20
*Sibbaldia procumbens*, 152
*Sidalcea glaucescens*, 108
    *malviflora*, 108, **110**
    *oregana spicata*, 108
*Silene californica*, 59, **60**
    *gallica*, 60
    *hookeri*, 59
*Silybum marianum*, 35
*Sisyrinchium bellum*, 94, **95**
    *californicum*, 94
*Smilacina*, 104
    *racemosa*, **105**
*Smilax californica*, 104
Solanaceae, 168–169
*Solanum nigrum*, 168
    *umbelliferum*, 168, **169**
    *xantii*, 168
*Solidago*, 35
*Sonchus*, 35
*Spartina foliosa*, 129
*Spartium junceum*, 82
*Spergula arvensis*, 60
*Spergularia*, 59
*Sphaeralcea ambigua*, 108, 109
*Spiranthes*, 119
*Sporobolus*, 128
*Stachys*, 98
    *rigida*, **100**
*Stanleya*, 56

*Stellaria media*, 59
*Stenanthium occidentale*, 104
*Stephanomeria*, 35
Sterculiaceae, 108
*Stillingia*, 80
*Stipa*, 128
*Streptanthus tortuosus*, 48, **50**
*Streptopus amplexifolius*, 104
*Stylomecon heterophylla*, 123
*Suaeda*, 61
*Swertia radiatum*, 86, **87**
*Symphoricarpos albus laevigatus*, 57, **58**
   *mollis*, 57
*Symphytum officinale*, 46
*Syringa*, 115
*Syzygium aromaticum*, 111

*Taraxacum officinale*, 35
*Tauschia*, 25
Taxaceae, 170–171
Taxodiaceae, 172
*Taxus brevifolia*, 170
*Tectona grandis*, 175
*Tellima grandiflora*, 162
*Tetracoccus*, 79
*Thalictrum*, 147
*Thamnosma montana*, 158
Themidaceae, 173–174
*Theobroma cacao*, 108
*Thermopsis macrophylla*, 82
*Thuja plicata*, 70
*Thysanocarpus curvipes*, 49, **50**
*Tiarella trifoliata unifoliata*, 162
*Tofieldia glutinosa*, 104
Tofieldiaceae, 104
*Tolmiea menziesii*, 162
*Torilis scandicina*, 25
*Torreya californica*, 170, **171**
*Toxicodendron diversilobum*, 22, **23**
   *radicans*, 22
*Trachelospermum jasminoides*, 28
*Tragia stylaris*, 80
*Trianthema portulacastrum*, 20
*Trichostema lanatum*, 98
  *parishii*, 98
*Trientalis latifolia*, 143
*Trifolium*, 82
Trilliaceae, 104
*Trillium*, 104
*Triodanis*, 54
*Triphysaria eriantha*, 166
*Triteleia*, 173
   *hyacinthina*, 174
   *ixioides*, 174
   *laxa*, 174

*Trollius*, 147
*Tsuga heterophylla*, 125
   *mertensiana*, 125
*Tulipa*, 103
*Turricula parryi*, 92

*Ulex europea*, 82
*Umbellularia californica*, 101, **102**
*Uropappus lindleyi*, 35

*Vaccinium ovatum*, 75
   *parvifolium*, 75
*Vancouveria planipetala*, 42, **43**
*Vanda*, 119
*Veratrum*, 104
*Verbacsum*, 165
*Verbena gooddingii*, 175
   *hastata*, 175, **176**
   *lasiostachys*, 175
Verbenaceae, 175–176
*Veronica*, 165
Veronicaceae, 165
*Viburnum ellipticum*, 57
*Vicia americana*, 82
   *gigantea*, 82
*Vinca*, 28
   *major*, 29, **31**
*Viola adunca*, 177
   *douglasii*, 177
   *glabella*, 177
   *lobata*, 177
   *macloskeyi*, 177
   *ocellata*, 177
   *pedunculata*, 177
   *sempervirens*, 177, **178**
   *sheltonii*, 177
Violaceae, 177–78
Vitaceae, 68

*Washingtonia filifera*, 32, **33**
   *robusta*, 32
*Watsonia*, 94
*Wislizenia refracta*, 56
*Wyethia helenioides*, 36, **39**

*Xanthium*, 36
*Xerophyllum tenax*, 105

*Yucca baccata*, 18
   *brevifolia*, 18
   *schidigera*, 18

*Zigadenus*, 104
*Ziziphus jujuba*, 150
*Zygocactus*, 51

# ABOUT THE AUTHOR AND ILLUSTRATOR

### GLENN KEATOR

Glenn Keator is a botanist, teacher, and writer specializing in California native plants. He teaches at several venues around the San Francisco Bay Area. He has written several books, including two other University of California Press titles: *Introduction to Trees of the San Francisco Bay Region* and *Designing California Native Gardens* (coauthored with Alrie Middlebrook).

### MARGARET J. STEUNENBERG

Margaret J. Steunenberg is a natural science illustrator, and the focus of her work is education. In addition to this publication, her artwork has appeared in *Bay Nature* magazine, *Plants of the East Bay Parks* (Roberts Rinehart), and *Today's Botanical Artists* (Schiffer Publishing). Her work has been in juried exhibits at the Smithsonian Institution, the Missouri Botanical Garden, and the New York State Museum. She has been honored with a Gold Award from the San Francisco Society of Illustrators.